International Dictionary of Refrigeration

Diccionario Internacional del Frío

Dictionnaire International du Froid

International Institute of Refrigeration (IIR)
Institut International du Froid (IIF)

International Dictionary of Refrigeration
Diccionario Internacional del Frío
Dictionnaire International du Froid

SPANISH-ENGLISH-FRENCH: Terms

ESPAÑOL-INGLÉS-FRANCÉS: términos

ESPAGNOL-ANGLAIS-FRANÇAIS: termes

Editor / Editeur scientifique: IIF-IIR

PEETERS

2008

International Institute of Refrigeration
Institut International du Froid
IIF-IIR, 177 boulevard Malesherbes, 75017 Paris, France
E-Mail: iif-iir@iifiir.org

PEETERS

ISBN: 978-90-429-2134-4 (Peeters Leuven)
ISBN: 978-2-7584-0049-3 (Peeters France)

D. 2008/0602/121

ÍNDICE

PREFACIO

El *Diccionario Internacional del Frío* es un conjunto de volúmenes integrado por un volumen básico *inglés-francés* con los términos y sus definiciones, más nueve volúmenes cada uno de los cuales contiene solamente los términos, así como su equivalente en inglés y en francés, en uno de los siguientes idiomas: alemán, árabe, chino, **español** (el presente volumen), holandés, italiano, japonés, noruego y ruso.

Una vez publicados todos los volúmenes, se prevé la publicación de un CD-ROM interactivo, así como la de un volumen que agrupe todos los idiomas citados.

El Diccionario contiene 4.000 términos aproximadamente, de los cuales el 20% son sinónimos, lo que equivale a dar unas 3500 definiciones en el volumen *inglés-francés*; un avance si se compara con los 3000 términos y 2400 definiciones de la obra publicada por el Instituto Internacional del Frío (IIF) en 1975.

El Diccionario refleja la diversidad de países miembros del IIF, cuyos dos idiomas oficiales son el inglés y el francés. En relación a la obra de 1975 se han añadido el árabe, el chino y el japonés.

También refleja la diversidad y la evolución de las aplicaciones del frío, con muchos términos nuevos especialmente en los campos del acondicionamiento de aire, la criobiología, los fluidos frigorígenos y los gases licuados. Incluye dos capítulos totalmente nuevos dedicados a las bombas de calor y al medio ambiente. La mayor parte de las definiciones de los términos existentes ha sido revisada y se han eliminado más de doscientos términos que han quedado obsoletos.

La elaboración del Diccionario ha tenido lugar entre los años 2000-2007, con las siguientes etapas importantes:

- en 2000 se seleccionaron las palabras clave y se redactó un resumen de la obra y una tabla de correspondencias entre los capítulos del Diccionario y las comisiones científicas y tecnológicas del IIF. Paralelamente se llevó la totalidad de la obra de 1975 a formato informático.

  Durante ese trabajo de preparación preliminar nos hemos inspirado en las recomendaciones terminológicas establecidas por distintas organizaciones internacionales, en especial por la Organización Internacional de Normalización (ISO), por determinados organismos europeos y/o americanos afiliados o no a ISO, así como por distintos órganos de Naciones Unidas;

- en 2001 y 2002 se formaron diez grupos técnicos de trabajo con participantes voluntarios, correspondientes a las diez comisiones del IIF, cuya tarea consistió en actualizar los términos que componían el Diccionario. El idioma de trabajo fue el inglés;

- en 2002 y 2003 esos mismos grupos trabajaron en las definiciones en inglés de aquellos términos;

- de 2003 a 2005 diez grupos de traducción se encargaron de traducir los términos a los idiomas elegidos. Entre aquellos se incluye el grupo francés, que también tradujo las definiciones al francés, y el grupo español. Generalmente fueron las asociaciones nacionales del frío de los países interesados las que se hicieron cargo de la coordinación de los trabajos, con las siguientes excepciones:

  - para el árabe, el delegado del Líbano en el Comité Ejecutivo del IIF, en colaboración con los delegados de siete países de lengua árabe miembros del IIF;

- para el español, el Instituto del Frío;
- para el ruso, el Comité nacional ruso para el IIF.

  - de 2005 a 2006 el trabajo de los grupos técnicos fue revisado por los presidentes de Comisión;
  - en 2006 y 2007 se realizó la revisión final del Diccionario con vistas a su publicación. Los grupos de traducción intervinieron de nuevo para traducir los 300 términos nuevos aparecidos durante la elaboración del cuerpo de la obra, así como la validación del conjunto de los términos.

Cerca de 200 expertos miembros de la red del IIF y pertenecientes a una treintena de países de todos los continentes han participado en la elaboración del Diccionario. Desde aquí agradecemos efusivamente su trabajo así como el de las personas de la sede del IIF que han colaborado, en especial: Daniel Viard, Director Adjunto del IIF y jefe del proyecto; Dany Furteau; Agnès Josserand-Broise y Susan Phalippou Mitchell.

El Diccionario tiene por objeto responder a las necesidades de un amplio abanico de instituciones y de usuarios, en especial: universidades (departamentos de ingeniería mecánica, eléctrica, química, energética ...), investigadores en el campo del frío y del acondicionamiento de aire, ingenieros, técnicos, servicios de información de las empresas, y consultores.

Agradecemos igualmente a Ediciones Peeters el haber aceptado el encargo de publicar el Diccionario.

Finalmente, partiendo del cuerpo de la obra así constituido prevemos la aparición regular de reediciones que tendrán en cuenta la dinámica de la evolución del campo del frío, junto con la posibilidad de incorporar nuevos idiomas.

El Diccionario Internacional del Frío ambiciona convertirse en una herramienta de trabajo indispensable para todos aquellos que trabajan en los diferentes oficios del frío. Esperamos que nuestros lectores nos hagan llegar sus comentarios para que cada día sea más útil.

Didier Coulomb, Director del IIF

AGRADECIMIENTOS

RELACION DE EXPERTOS QUE HAN INTERVENIDO EN LA PREPARACION DEL *DICCIONARIO INTERNA-CIONAL DEL FRIO*

Los 10 **grupos temáticos de trabajo**, correspondientes a las Comisiones del IIF (los revisores son los Presidentes de Comisión en ejercicio) y que han utilizado el idioma inglés en su trabajo:

Comisión A1: Criofísica, criotécnica

Ivan A. Arkharov	Rusia
Guobang Chen	China
Guy Gistau Baguer – *coordinador*	Francia
Philippe Lebrun – *revisor*	Francia
Rachid Rebiai	Argelia

Comisión A2: Licuefacción y separación de gases

Rakesh Agrawal	Estados Unidos
Walter Castle – *coordinador*	Reino Unido
Václav Chrz – *revisor*	República Checa
Cyril Collins	Reino Unido
Wolfgang Foerg	Alemania
Krish R. Krishnamurthy	Estados Unidos
Sebastian Muschelknautz	Alemania
Rachid Rebiai	Argelia
Vadim Udut	Rusia

Comisión B1: Termodinámica y procesos de transferencia

Pradeep K. Bansal	Nueva Zelanda
Jocelyn Bonjour	Francia
Stefan Ditchev	Bulgaria
Piotr Domanski – *coordinador*	Estados Unidos
Carlos Alberto Infante Ferreira	Holanda
Dieter Gorenflo – *revisor*	Alemania
Josef Ota	República Checa
Milan Šafr	República Checa
Oleg B. Tsvetkov	Rusia
Koichi Watanabe	Japón
Felix Ziegler	Alemania

Comisión B2: Equipos frigoríficos

Ben Adamson	Australia
Clark Bullard – *revisor*	Estados Unidos

Carmine Casale	Italia
Mario Costantino	Italia
Daniele Farina	Italia
Ezio Fornasieri	Italia
Michael Kauffeld	Alemania
Grégoire Lambrinos	Grecia
Siegfried Nowotny – *coordinador*	Alemania
Natividad López Rodriguez	España
Filippo de Rossi	Italia
Milan Šafr	República Checa
Wolfgang Sussek	Alemania
Josef Weiss	Austria
Felix Ziegler	Alemania

Comisión C1: Criobiología, criomedicina

Anne Baudot	Francia
Jean-Paul Homasson – *revisor*	Francia
Norio Murase	Japón
Andreas Sputtek	Alemania

Comisión C2: Ciencia e ingenieria alimentarias

Bryan Becker – *co-coordinador*	Estados Unidos
Leif Bogh Sorensen	Dinamarca
Christopher John Kennedy – *co-coordinador*	Reino Unido
Grégoire Lambrinos	Grecia
Alain Le Bail	Francia
Bart Nicolaï – *revisor*	Belgica
José Luis de la Plaza	España
Rikuo Takai	Japón
Josef Weiss	Austria

Comisión D1: Almacenamiento frigorífico

Milan Houska – *coordinador*	República Checa
Manuel Lamúa Soldevilla	España
Anders Lindborg	Suecia
Eleni Manolopoulou	Grecia
Marek Michniewicz	Polonia
David J. Tanner – *revisor*	Nueva Zelanda

Comisión D2: Transporte frigorífico

Nevin Amos – *co-coordinador*	Nueva Zelanda
Robert Heap – *co-coordinador*	Reino Unido
Sung Lim Kwon	Estados Unidos
Hisashi Mitsufuji	Japón
Girolamo Panozzo – *revisor*	Italia
Rikuo Takai	Japón
David Tanner	Nueva Zelanda
Ruhe Xie	China

Comisión E1: Acondicionamiento de aire

Karel Hemzal	República Checa
Peter Novak	Eslovenia
Weiding Long – *revisor*	China
R. E. «Sam» Luxton	Australia
Pietro Mazzei – *coordinador* (términos)	Italia
Jaroslav Wurm† – *coordinador* (definiciones)	Estados Unidos

Comisión E2: Bombas de calor, recuperación de energía

Jos Bouma – *coordinador*	Holanda
Alberto Coronas Salcedo	España
Daniele Farina	Italia
Hermann Halozan – *revisor*	Autria
Václav Havelský	Eslovaquia
Ahmed Kamoun	Túnez
Petter Nekså	Noruega
S. Srinivasa Murthy – *vice- coordinador*	India

También se han formado 10 **grupos de idiomas**.

Los miembros del grupo de trabajo para el español, organizados por el Instituto del Frío de Madrid, han sido:

Francisco Artés Calero
Alejandro Cabetas
José Miguel Díaz Serrano – *coordinador y revisor*
Jesús Espinosa
Josefa Fernández
Manuel Lamúa Soldevilla
Antonio Molina
José Luis de la Plaza

Queremos agradecer vivamente a todos nuestros colegas españoles del grupo de trabajo, así como al Instituto del Frío de Madrid, la considerable tarea que han llevado a cabo.

Los miembros del grupo de trabajo para el idioma francés, en colaboración con la Asociación Francesa del Frío (AFF), han sido:

Mohamed Salah Abid (Túnez)	Jean-Paul Homasson
Anne Baudot	Alain Le Bail
Jean-Pierre Besancenot	Philippe Lebrun
Jocelyn Bonjour	Pierre Leclère – *vice-coordinador*
François Clément – *coordinador*	Bernard Lelièvre
Félix Depledt	Jean Rémy
Claude Dessaux	Gérard Renaudin
Maxime Duminil – *vice-coordinador*	Bernard Saulnier
Claude Ernst	Maurice Serrand
Guy Gistau Baguer	Georges Vrinat
Alain Goy	

NOTA A LOS LECTORES

Con el fin de facilitar la utilización del volumen de términos *Español-inglés-francés* del *Diccionario Internacional del Frío,* se han adoptado las normas de presentación que figuran a continuación:

1. El núcleo del volumen – con los términos únicamente – está organizado en primer lugar según una clasificación temática, y alfabética después.

El volumen *Español-inglés-francés* comprende unos 4.400 términos que son palabras o expresiones formadas por varias palabras, en los tres idiomas, siendo sinónimos el 20% de los términos, aproximadamente.

Los términos de la obra se agrupan en 11 capítulos que pueden estar subdivididos en varios niveles jerárquicos de subcapítulos. Tan solo los Capítulos 8 y 11 carecen de subdivisiones. Dentro de cada uno de los 110 subcapítulos los términos se clasifican por orden alfabético.

1.1. Organización y jerarquización

Dentro de las últimas subdivisiones jerárquicas los términos se clasifican por el orden alfabético de los términos ingleses, situándose los términos españoles entre los ingleses y franceses correspondientes.

Todos los términos ingleses, escritos en caracteres estándar, figuran en este orden. Cada término lleva asociado un código que se encuentra en la columna situada a la derecha de la hoja en las páginas impares, y a la izquierda en las páginas pares. El código está formado por el número del subcapítulo seguido de un guión y el número de orden alfabético del término en dicho subcapítulo. Este último número aparece en **negrita**: por ejemplo, 1.1.4-**3**.

Cada término va acompañado automáticamente de sus posibles sinónimos escritos en *cursiva*. En cada grupo de términos sinónimos uno de ellos ha sido elegido como término principal. Cuando el término principal figura como sinónimo de otros términos del grupo, se escribe en *cursiva* y va siempre delante de los demás sinónimos.

En el volumen *inglés-francés* la definición en inglés que corresponde al término principal y a sus sinónimos aparece junto al término principal. Frente al término principal inglés homólogo, el término principal francés y sus sinónimos van acompañados de la definición francesa correspondiente a la definición inglesa.

Los términos en español, acompañados o no de uno o varios sinónimos, no están en orden alfabético. Sin embargo, se puede uno remitir al índice alfabético de los términos españoles.

El término principal español en **negrita**, junto con sus sinónimos escritos en caracteres estándar, aparece al lado de cada término de un grupo de sinónimos ingleses. También aparece el término principal francés en caracteres estándar acompañado de sus sinónimos en *cursiva*.

Existen unos cuarenta términos ingleses (y franceses) con suficiente importancia como para figurar en dos subcapítulos diferentes. El código adoptado en el índice español corresponde al código del término principal inglés cuya definición figura en uno de los dos subcapítulo del volumen *inglés-francés*.

1.2. Presentación y topografía

– Se han considerado tres categorías de términos en inglés y en francés:
  • los términos preferidos;
  • los términos aceptados (de uso corriente);
  • los términos obsoletos (susceptibles de ser utilizados todavía)

Como las dos últimas categorías representan algo menos del 15% de los términos en inglés (el 11% y el 3%, respectivamente), se ha optado por poner las siguientes señales al lado de cada uno de ellos:

  ◐ término aceptado
  ○ término obsoleto

Cuando un término se asocia a sinónimos que pertenecen a varias categorías, éstos aparecen en el orden siguiente: en primer lugar los términos preferidos, seguidos de los términos aceptados, y en última posición los términos obsoletos.

Esta distinción no se hace con los términos españoles; solo se diferencia entre el término principal, en **negrita**, y el o los sinónimos, en caracteres estándar.

Los siguientes ejemplos ilustran los principios enunciados anteriormente:

boiler ○	**hervidor** (subst.)	générateur	2.3.3.1-**12**
generator	generador (subst.)	*bouilleur* ○	
regenerator			
desorber ◐			

desorber ◐	**hervidor** (subst.)	générateur	2.3.3.1-**13**
generator	generador (subst.)	*bouilleur* ○	
regenerator			
boiler ○			

generator	**hervidor** (subst.)	générateur	2.3.3.1-**19**
regenerator	generador (subst.)	*bouilleur* ○	
desorber ◐			
boiler ○			

regenerator	**hervidor** (subst.)	générateur	2.3.3.1-**27**
generator	generador (subst.)	*bouilleur* ○	
desorber ◐			
boiler ○			

– El código del término principal inglés se encuentra, por si fuera necesario, en el índice alfabético español.

– Algunos términos españoles, ingleses y franceses pueden estar seguidos de : (adj.) o (subst.):
  adj.   = adjetivo (en español), adjective (en inglés) o adjectif (en francés),
  subst. = substantivo (en español), substantive (en inglés) o substantif (en francés).

– Si algún término no tiene equivalente en un idioma, en la columna correspondiente aparecen tres asteriscos ***.

2. Índice alfabético de términos en español:

–   Cada término español va seguido del código del término principal inglés correpondiente, cuya definición se encuentra en el volumen *inglés-francés*.

–   Un mismo término español puede tener varios significados, en cuyo caso puede ir seguido de varios códigos pertenecientes a los términos principales en inglés correspondientes.

Para tener acceso a las definiciones de los términos españoles en inglés y en francés es indispensable remitirse al volumen *inglés-francés*, que es el volumen básico del *Diccionario Internacional del Frío*.

NOTES FOR READERS

In order to facilitate the use of the *Spanish-English-French* volume of the *International Dictionary of Refrigeration,* the following rationale was used in the presentation:

1. The core of the volume – containing the terms only – is organized primarily according to themes and is in alphabetical order.

The *Spanish-English-French* volume comprises approximately 4400 terms that are words or expressions composed of several words in the three languages; 20% of the terms are synonyms.

The terms are allocated to 11 chapters that are each subdivided into several sections of various levels of importance (only Chapters 8 and 11 are not subdivided). Within the 110 sections, the terms are classified in alphabetical order.

1.1. Organization and hierarchic classification

Under the smallest headings in the hierarchy, the terms are classified according to the alphabetical order of the English terms, and the Spanish terms are placed between the related English and French terms.

All the English terms in standard prints are arranged in this manner.
A code is assigned to each term and is placed on the right of odd-numbered pages and on the left of even-numbered pages. It is composed of the section number, separated by a dash from the rank of the term in the alphabetical order, within that particular section, and the latter number appears in **bold**: for example, 1.1.4-**3.**

For each term, any synonym appears in *italics.*
In each group of synonymous terms, one term has been selected as the main term.
When the main term appears as a synonym of one of the terms in the group, it appears in *italics,* again in first position, before the other synonyms.

In the *English-French* volume, the English definition referring to the main term and to its synonyms is mentioned with the main term and, opposite the main synonymous English term, the main French term and its synonyms are provided with the French definition relating to the English definition.

The Spanish terms, with or without one or several synonyms, are not arranged in alphabetical order. However, one can refer to the alphabetical index of Spanish terms.

Opposite each term in a group of English synonyms, the main Spanish term, in **bold,** is given along with all its synonyms, in standard print, and the main French term is in standard print, along with its synonyms in *italics.*

About 40 English (and French) terms were deemed sufficiently important to be placed in two different sections. The code figuring in the Spanish index is related to the main English term whose definition is to be found in one of these two sections of the *English-French* volume.

1.2. Presentation and typography

– We divided terms in English and French into three categories:
  * preferred terms;
  * tolerated terms (commonly used);
  * outdated terms (that may still be used).

The last two categories cover slightly less than 15% of the English terms (11% and 3% respectively) and we decided to mention the following after the terms involved:

  ◐ tolerated term
  ○ outdated term

When a term has synonyms in several categories, these categories are presented in the following order: preferred terms first, tolerated terms in second position and finally the outdated terms.

This distinction was not made in the case of Spanish terms; there is only a distinction between the main term in **bold** print and the synonym(s) in standard print.

Here is an example illustrating the principles described above:

boiler ○ *generator* *regenerator* *desorber* ◐	**hervidor** (subst.) generador (subst.)	générateur *bouilleur* ○	2.3.3.1-**12**
desorber ◐ *generator* *regenerator* *boiler* ○	**hervidor** (subst.) generador (subst.)	générateur *bouilleur* ○	2.3.3.1-**13**
generator *regenerator* *desorber* ◐ *boiler* ○	**hervidor** (subst.) generador (subst.)	générateur *bouilleur* ○	2.3.3.1-**19**
regenerator *generator* *desorber* ◐ *boiler* ○	**hervidor** (subst.) generador (subst.)	générateur *bouilleur* ○	2.3.3.1-**27**

– If necessary, the code of the main English term can be found in the Spanish alphabetical index.

– For certain terms, (adj.) or (subst.) are added:

  adj.   = adjetivo (in Spanish), adjective (in English) or adjectif (in French)
  subst. = substantive (in Spanish), substantive (in English) or substantif (in French)

– If a term has no equivalent in a language, three asterisks*** are placed in the corresponding column.

2. Alphabetical Index of the Spanish terms:

– Each Spanish term is accompanied with the corresponding code of the main related English term, leading directly to the definition in the *English-French* volume.

- A same Spanish term may have several meanings, in which case it can be accompanied with several codes, namely those of the related main English terms.

In order to access the definitions of the Spanish terms in English and French, it is essential to refer to the main *English-French* volume of the *International Dictionary of Refrigeration*.

AVIS AU LECTEUR

Afin de faciliter l'utilisation du volume des termes *Espagnol*-*anglais-français* du *Dictionnaire International du Froid*, les principes de présentation suivants ont été adoptés :

1. Le cœur du volume – avec les termes seulement – est organisé selon un classement thématique, puis alphabétique.

Le volume *Espagnol*-*anglais-français* est composé d'environ 4400 termes, qui sont des mots ou des expressions formées de plusieurs mots, dans les trois langues, 20 % des termes étant des synonymes.

Les termes de l'ouvrage sont regroupés dans 11 chapitres, qui peuvent être eux-mêmes subdivisés en plusieurs niveaux hiérarchiques de sous-chapitres (seuls les Chapitres 8 et 11 ne sont pas subdivisés). Au sein des 110 sous-chapitres, les termes sont classés par ordre alphabétique.

1.1. Organisation et hiérarchisation

A l'intérieur des plus petites subdivisions hiérarchiques, les termes sont classés selon l'ordre alphabétique des termes anglais, les termes espagnols étant situés entre les termes anglais et français correspondants.

Tous les termes anglais, marqués en caractères standards, défilent dans cet ordre.
A chaque terme est associé un code qui figure dans la colonne située à droite sur les pages impaires et à gauche sur les pages paires. Il est constitué du numéro du sous-chapitre séparé par un trait d'union du numéro d'ordre alphabétique du terme dans le sous-chapitre, ce dernier numéro étant écrit en **gras** : par exemple, 1.1.4-**3**.

Chacun des termes est systématiquement accompagné de ses synonymes éventuels marqués en *italique*.
Dans chaque groupe de termes synonymes, l'un d'entre eux a été choisi comme terme principal. Lorsque le terme principal est en position de synonyme par rapport à un autre terme du groupe, il est écrit en *italique*, toujours en tête des autres synonymes.

Dans le volume *anglais-français*, la définition anglaise correspondant au terme principal et à ses synonymes est mentionnée avec le terme principal et, en face du terme principal anglais homologue, le terme principal français et ses synonymes sont accompagnés de la définition française correspondant à la définition anglaise.

Les termes espagnols, accompagnés ou non d'un ou de plusieurs synonymes, ne sont pas dans l'ordre alphabétique. Mais on peut se référer à l'index alphabétique des termes espagnols.

En face de chacun des termes d'un groupe de synonymes anglais, on retrouve le terme principal espagnol en **gras**, accompagné de l'ensemble de ses synonymes en caractères standards, et le terme principal français, en caractères standards, accompagné de ses synonymes en *italique*.

Une quarantaine de termes anglais (et français) ont été considérés comme suffisamment importants pour être présentés dans deux sous-chapitres différents. Le code retenu dans l'index espagnol correspond au code du terme anglais principal dont on a mis la définition dans l'un des deux sous-chapitres du volume *anglais-français*.

1.2. Présentation et typographie

– Nous avons distingué trois catégories de termes en anglais et en français :
  - les termes préférables,
  - les termes tolérés (d'usage courant),
  - les termes désuets (susceptibles d'être encore utilisés).

Les deux dernières catégories représentant un peu moins de 15 % des termes anglais (respectivement 11 % et 3 %), nous avons choisi d'adjoindre la marque suivante après chacun des termes concernés :

  ◐ terme toléré
  ○ terme désuet

Lorsqu'un terme est associé à des synonymes appartenant à plusieurs catégories, ceux-ci sont présentés dans l'ordre suivant : les termes préférables en premier, les termes tolérés en deuxième et les termes désuets en dernier.

Cette distinction n'a pas été faite avec les termes espagnols ; il y a seulement la distinction entre le terme principal en **gras** et le (ou les) synonyme(s) en caractères standards.

Exemple illustrant les principes énoncés précédemment :

boiler ○ *generator* *regenerator* *desorber* ◐	**hervidor** (subst.) generador (subst.)	générateur *bouilleur* ○	2.3.3.1-**12**
desorber ◐ *generator* *regenerator* *boiler* ○	**hervidor** (subst.) generador (subst.)	générateur *bouilleur* ○	2.3.3.1-**13**
generator *regenerator* *desorber* ◐ *boiler* ○	**hervidor** (subst.) generador (subst.)	générateur *bouilleur* ○	2.3.3.1-**19**
regenerator *generator* *desorber* ◐ *boiler* ○	**hervidor** (subst.) generador (subst.)	générateur *bouilleur* ○	2.3.3.1-**27**

– Si nécessaire, on trouvera dans l'index alphabétique espagnol le code du terme principal anglais ;

– Certains termes espagnols, anglais et français peuvent être accompagnés de la mention : (adj.) ou (subst.) ;

  adj.   = adjetivo (en espagnol), adjective (en anglais) ou adjectif (en français)
  subst. = substantivo (en espagnol), substantive (en anglais) ou substantif (en français) ;

– Si un terme n'a pas d'équivalent dans une langue, trois astérisques *** sont placées dans la colonne correspondante.

2. Index alphabétique des termes espagnols :

– Chaque terme espagnol est accompagné du code du terme principal anglais correspondant, dont on a la définition dans le volume *anglais-français*.

– Un même terme espagnol peut avoir plusieurs significations, auquel cas il peut être accompagné de plusieurs codes, ceux des termes principaux anglais correspondants.

Pour avoir accès aux définitions des termes espagnols en anglais et en français, il est indispensable de se référer au volume *anglais-français*, volume de base du *Dictionnaire International du Froid*.

ESTRUCTURA GENERAL

Como consecuencia del abanico de idiomas que se incluyen, el *Diccionario Internacional del Frío* está integrado por:

- un volumen básico *inglés-francés* que consta de un núcleo de términos con sus definiciones, seguido de un índice alfabético de los términos en inglés y en francés.

- Nueve volúmenes complementarios, uno para cada idioma, con los términos únicamente, cada uno de los cuales se relaciona con su equivalente en inglés y en francés.
  - alemán, con un índice en alemán,
  - árabe, con un índice en inglés y en francés,
  - chino, con un índice en inglés,
  - **español**, con un índice en español,
  - holandés, con un índice en holandés,
  - italiano, con un índice en italiano,
  - japonés, con un índice en inglés,
  - noruego, con un índice en noruego,
  - ruso, con un índice en ruso.

Una vez que todos los volúmenes hayan sido publicados, se prevé la publicación de un CD-ROM interactivo y la de un volumen único que agrupe todos los idiomas.

Capítulo 1. | FUNDAMENTOS

◑ término aceptado

○ término obsoleto

ENGLISH	ESPAÑOL	FRANÇAIS	
SECTION 1.1 *General concepts and instrumentation* **SECTION 1.1.1** *General concepts and instrumentation:* *general background*	**SUBCAPÍTULO 1.1** *Conceptos generales e instrumentación* **SUBCAPÍTULO 1.1.1** *Generalidades sobre conceptos* *generales e instrumentación*	**SOUS-CHAPITRE 1.1** *Concepts généraux et instrumentation* **SOUS-CHAPITRE 1.1.1** *Généralités sur les concepts généraux* *et l'instrumentation*	
authorized person	**persona autorizada**	personne autorisée	1.1.1-**1**
British thermal unit (Btu) ◐	**BTU (unidad térmica inglesa)**	Btu ◐	1.1.1-**2**
cold (adj.)	**frío** (adj.)	froid (adj.)	1.1.1-**3**
cold chain	**cadena del frío** cadena frigorífica	chaîne du froid	1.1.1-**4**
competence	**capacidad** competencia	compétence	1.1.1-**5**
congeal (to) *freeze (to)*	**congelar**	geler	1.1.1-**6**
cooling	**enfriamiento**	refroidissement	1.1.1-**7**
critical density	**densidad crítica** densidad en condiciones críticas	masse volumique critique	1.1.1-**8**
density	**densidad** peso específico masa específico	masse volumique	1.1.1-**9**
food chain	**cadena alimentaria** producción y procesado de alimentos	chaîne alimentaire	1.1.1-**10**
freeze (to) *congeal (to)*	**congelar**	geler	1.1.1-**11**
frost	**hielo** escarcha	givre	1.1.1-**12**
(graduate) refrigerating engineer ◐ *(graduate) refrigeration engineer*	**ingeniero frigorista**	ingénieur frigoriste	1.1.1-**13**
(graduate) refrigeration engineer *(graduate) refrigerating engineer* ◐	**ingeniero frigorista**	ingénieur frigoriste	1.1.1-**14**
heat capacity *thermal capacity*	**capacidad calorífica** capacidad térmica	capacité thermique	1.1.1-**15**
joule (J)	**julio (J)**	joule (J)	1.1.1-**16**
kilocalorie (kcal) ◐	**kilocaloría (kcal)** caloría grande	kilocalorie (kcal) ◐ *millithermie (mth)* ○	1.1.1-**17**
newton (N)	**newton (N)**	newton (N)	1.1.1-**18**
pascal (Pa)	**pascal (Pa)**	pascal (Pa)	1.1.1-**19**
reference value	**valor de referencia**	valeur de référence	1.1.1-**20**
refrigerate (to)	**enfriar (artificialmente)** refrigerar	refroidir	1.1.1-**21**
refrigerating	**frigorífico** (adj.)	frigorifique (adj.)	1.1.1-**22**
refrigerating capacity	**potencia frigorífica** potencia de enfriamiento	puissance frigorifique	1.1.1-**23**
refrigeration	**enfriamiento (artificial)** frío artificial	froid (artificiel)	1.1.1-**24**
refrigeration contractor	**instalador frigorista**	entrepreneur frigoriste	1.1.1-**25**
refrigeration engineer ◐ *refrigeration mechanic*	**mecánico frigorista**	mécanicien frigoriste	1.1.1-**26**

		ENGLISH	ESPAÑOL	FRANÇAIS
1.1.1-**27**		retrigeration engineering	**técnica frigorífica** técnica del frío ingeniería del frío	génie frigorifique *technique frigorifique* *technique du froid*
1.1.1-**28**		refrigeration installation engineer	**montador frigorista**	monteur frigoriste
1.1.1-**29**		refrigeration mechanic *refrigeration engineer* ⊙	**mecánico frigorista**	mécanicien frigoriste
1.1.1-**30**		refrigeration serviceman *refrigeration technician* ⊙	**reparador frigorista** técnico frigorista	dépanneur frigoriste
1.1.1-**31**		refrigeration technician ⊙ *refrigeration serviceman*	**reparador frigorista** técnico frigorista	dépanneur frigoriste
1.1.1-**32**		relative molecular mass	**masa molecular relativa**	masse moléculaire relative
1.1.1-**33**		specific gravity ⊙	**densidad**	densité ⊙
1.1.1-**34**		specific heat c *specific heat capacity c*	**calor específico** calor másico	capacité thermique massique c *chaleur massique c* ○ *chaleur spécifique c* ○
1.1.1-**35**		specific heat capacity c *specific heat c*	**calor específico** calor másico	capacité thermique massique c *chaleur massique c* ○ *chaleur spécifique c* ○
1.1.1-**36**		specific heat capacity c_p at constant pressure	**capacidad calorífica específica a presión constante** calor específico a presión constante	capacité thermique massique c_p à pression constante *chaleur massique à pression constante c_p* ○ *chaleur spécifique à pression constante c_p* ○
1.1.1-**37**		specific heat capacity c_v at constant volume	**capacidad calorífica específica a volumen constante** calor específico a volumen constante	capacité thermique massique c_v à volume constant *chaleur massique à volume constant c_v* ○ *chaleur spécifique à volume constant c_v* ○
1.1.1-**38**		specific heat ratio c_p/c_v	**coeficiente adiabático**	rapport des capacités thermiques massiques c_p/c_v
1.1.1-**39**		specific volume	**volumen específico** volumen másico	volume massique
1.1.1-**40**		therm ○	**termia (100 000 BTU)**	***
1.1.1-**41**		thermal capacity *heat capacity*	**capacidad calorífica** capacidad térmica	capacité thermique
1.1.1-**42**		ton of refrigeration (T.R.) ⊙	**toneladas de refrigeración (TR)**	***
1.1.1-**43**		watt (W)	**vatio (W)**	watt (W)
1.1.1-**44**		***	**frigoría** kilocaloría (negativa)	frigorie (fg) ○
1.1.1-**45**		***	**termia**	thermie (th) ○

		SECTION 1.1.2 *Temperature*	**SUBCAPÍTULO 1.1.2** *Temperatura*	**SOUS-CHAPITRE 1.1.2** *Température*
1.1.2-**1**		absolute temperature *thermodynamic temperature*	**temperatura absoluta** temperatura termodinámica	température absolue *température thermodynamique*
1.1.2-**2**		absolute zero	**cero absoluto**	zéro absolu
1.1.2-**3**		acoustical thermometer	**termómetro acústico**	thermomètre acoustique
1.1.2-**4**		alcohol thermometer	**termómetro de alcohol**	thermomètre à alcool
1.1.2-**5**		ambient temperature	**temperatura ambiente**	température ambiante
1.1.2-**6**		arithmetic mean temperature difference	**diferencia media aritmética de temperatura**	écart moyen arithmétique de température

ENGLISH	ESPAÑOL	FRANÇAIS	
average temperature *mean temperature*	**temperatura media**	température moyenne	1.1.2-**7**
bimetallic element *bimetallic strip*	**bimetal** elemento bimetálico bilámina	bilame	1.1.2-**8**
bimetallic strip *bimetallic element*	**bimetal** elemento bimetálico bilámina	bilame	1.1.2-**9**
bimetallic thermometer	**termómetro bimetálico** termómetro de bilámina	thermomètre à bilame	1.1.2-**10**
Celsius (temperature) scale	**escala (termométrica) Celsius** escala (termométrica) centígrada	échelle (thermométrique) Celsius	1.1.2-**11**
Celsius temperature	**temperatura Celsius**	température Celsius	1.1.2-**12**
Curie point	**punto de Curie**	point de Curie	1.1.2-**13**
degree Celsius (°C)	**grado Celsius (°C)** grado centígrado	degré Celsius (°C)	1.1.2-**14**
degree Fahrenheit (°F) ○	**grado Fahrenheit (°F)**	degré Fahrenheit (°F) ○	1.1.2-**15**
degree Rankine (°R) ○	**grado Rankine (°R)**	degré Rankine (°R) ○	1.1.2-**16**
dial thermometer	**termómetro de esfera** termómetro de cuadrante	thermomètre à cadran	1.1.2-**17**
distant reading thermometer ○ *remote reading thermometer*	**termómetro de lectura a distancia** termómetro a distancia teletermómetro termómetro remoto	téléthermomètre *thermomètre à distance* ○	1.1.2-**18**
electric contact thermometer	**termómetro de contactos eléctricos**	thermomètre à contacts électriques	1.1.2-**19**
external temperature *outside temperature*	**temperatura exterior**	température extérieure	1.1.2-**20**
Fahrenheit (temperature) scale ◉	**escala (termométrica) Fahrenheit**	échelle (thermométrique) Fahrenheit	1.1.2-**21**
gas thermometer	**termómetro de dilatación de gases**	thermomètre à gaz	1.1.2-**22**
indicating thermometer	**termómetro de lectura directa**	thermomètre à lecture directe	1.1.2-**23**
inside temperature *internal temperature*	**temperatura interior**	température intérieure	1.1.2-**24**
internal temperature *inside temperature*	**temperatura interior**	température intérieure	1.1.2-**25**
IPTS-68	**IPTS-68**	IPTS-68	1.1.2-**26**
ITS-90	**ITS-90**	ITS-90	1.1.2-**27**
kelvin (K)	**kelvin (K)**	kelvin (K)	1.1.2-**28**
Kelvin (temperature) scale *thermodynamic temperature scale*	**escala (termométrica) Kelvin**	échelle (thermométrique) Kelvin *échelle de température thermodynamique*	1.1.2-**29**
limiting temperature	**temperatura límite**	température limite	1.1.2-**30**
logarithmic mean temperature difference	**diferencia media logarítmica de temperatura**	écart moyen logarithmique de température	1.1.2-**31**
lowering of temperature *temperature reduction*	**disminución de la temperatura** descenso de temperatura	abaissement de température *réduction de température*	1.1.2-**32**
magnetic temperature	**temperatura magnética**	température magnétique	1.1.2-**33**
magnetic thermometer	**termómetro magnético**	thermomètre magnétique	1.1.2-**34**
maximum service temperature	**temperatura máxima de servicio** temperatura máxima de funcionamiento	température maximale de service	1.1.2-**35**

	ENGLISH	ESPAÑOL	FRANÇAIS
1.1.2-36	mean temperature *average temperature*	**temperatura media**	température moyenne
1.1.2-37	mean temperature difference	**diferencia media de temperatura**	écart moyen de température
1.1.2-38	mercury thermometer	**termómetro de mercurio**	thermomètre à mercure
1.1.2-39	minimum service temperature	**temperatura mínima de servicio** temperatura mínima de funcionamiento	température minimale de service
1.1.2-40	Néel point *Néel temperature* ●	**punto de Néel**	point de Néel
1.1.2-41	Néel temperature ● *Néel point*	**punto de Néel**	point de Néel
1.1.2-42	nuclear resonance thermometer	**termómetro de resonancia nuclear**	thermomètre à résonance nucléaire
1.1.2-43	operating temperature *working temperature* ●	**temperatura de trabajo** temperatura de funcionamiento	température de fonctionnement
1.1.2-44	outside temperature *external temperature*	**temperatura exterior**	température extérieure
1.1.2-45	pyrometer	**pirómetro**	pyromètre
1.1.2-46	Rankine temperature scale ●	**escala (termométrica) de Rankine**	échelle thermométrique Rankine ●
1.1.2-47	recording thermometer *temperature recorder* *thermograph* ●	**termógrafo** termómetro registrador	thermographe *thermomètre enregistreur*
1.1.2-48	reference mean temperature	**temperatura media de referencia**	température moyenne de référence
1.1.2-49	remote reading thermometer *distant reading thermometer* ○	**termómetro de lectura a distancia** termómetro a distancia teletermómetro termómetro remoto	téléthermomètre *thermomètre à distance* ○
1.1.2-50	resistance thermometer	**termómetro de resistencia**	thermomètre à résistance
1.1.2-51	solid expansion thermometer	**termómetro de dilatación de sólidos**	thermomètre à dilatation solide
1.1.2-52	standard ambient temperature	**temperatura ambiente normal** temperatura ambiente de referencia	température normale ambiante
1.1.2-53	subcooled refrigerant temperature	**temperatura de líquido frigorígeno subenfriado** temperatura de refrigerante subenfriado	température d'un fluide frigorigène sous-refroidi
1.1.2-54	surface temperature	**temperatura superficial** temperatura de la superficie	température de surface
1.1.2-55	temperature	**temperatura**	température
1.1.2-56	temperature difference	**diferencia de temperaturas** salto de temperatura	différence de température *écart de température*
1.1.2-57	temperature drop	**caída de temperatura**	chute de température
1.1.2-58	temperature fluctuation	**fluctuación de temperatura**	fluctuation de température
1.1.2-59	temperature gradient	**gradiente de temperatura**	gradient de température
1.1.2-60	temperature increment	**incremento de temperatura**	accroissement de température
1.1.2-61	temperature profile	**perfil de temperaturas**	profil des températures
1.1.2-62	temperature recorder *recording thermometer* *thermograph* ●	**termógrafo** termómetro registrador	thermographe *thermomètre enregistreur*
1.1.2-63	temperature reduction *lowering of temperature*	**disminución de la temperatura** descenso de temperatura	abaissement de température *réduction de température*
1.1.2-64	temperature rise	**elevación de (la) temperatura**	élévation de température

ENGLISH	ESPAÑOL	FRANÇAIS	
temperature scale	**escala de temperatura** escala termométrica	échelle de température *échelle thermométrique*	1.1.2-**65**
temperature variation	**variación de temperatura**	variation de température	1.1.2-**66**
thermal convector ○ *thermocouple* *thermo-junction* ◑ *thermoelectric couple* ◑	**par termoeléctrico** termopar	couple thermoélectrique *thermocouple*	1.1.2-**67**
thermistor	**termistor** termistancia	thermistance	1.1.2-**68**
thermo-junction ◑ *thermocouple* *thermoelectric couple* ◑ *thermal convector* ○	**par termoeléctrico** termopar	couple thermoélectrique *thermocouple*	1.1.2-**69**
thermocouple *thermo-junction* ◑ *thermoelectric couple* ◑ *thermal convector* ○	**par termoeléctrico** termopar	couple thermoélectrique *thermocouple*	1.1.2-**70**
thermodynamic temperature *absolute temperature*	**temperatura absoluta** temperatura termodinámica	température absolue *température thermodynamique*	1.1.2-**71**
thermodynamic temperature scale *Kelvin (temperature) scale*	**escala (termométrica) Kelvin**	échelle (thermométrique) Kelvin *échelle de température thermodynamique*	1.1.2-**72**
thermoelectric couple ◑ *thermocouple* *thermo-junction* ◑ *thermal convector* ○	**par termoeléctrico** termopar	couple thermoélectrique *thermocouple*	1.1.2-**73**
thermograph ◑ *recording thermometer* *temperature recorder*	**termógrafo** termómetro registrador	thermographe *thermomètre enregistreur*	1.1.2-**74**
thermometer	**termómetro**	thermomètre	1.1.2-**75**
thermometry	**termometría**	thermométrie	1.1.2-**76**
thermopile	**termopila**	thermopile	1.1.2-**77**
vapour pressure thermometer	**termómetro de tensión de vapor**	thermomètre à tension de vapeur	1.1.2-**78**
working temperature ◑ *operating temperature*	**temperatura de trabajo** temperatura de funcionamiento	température de fonctionnement	1.1.2-**79**

SECTION 1.1.3 *Pressure*	SUBCAPÍTULO 1.1.3 *Presión*	SOUS-CHAPITRE 1.1.3 *Pression*	
absolute pressure	**presión absoluta**	pression absolue	1.1.3-**1**
absolute pressure gauge	**medidor de presión absoluta** indicador de presión absoluta	manomètre absolu	1.1.3-**2**
absolute vacuum	**vacío absoluto**	vide absolu	1.1.3-**3**
annular chamber	**cámara anular**	chambre annulaire *espace annulaire*	1.1.3-**4**
atmospheric pressure	**presión atmosférica**	pression atmosphérique	1.1.3-**5**
barometer	**barómetro**	baromètre	1.1.3-**6**
bellows	**fuelle**	soufflet	1.1.3-**7**
Bourdon gauge	**manómetro de Bourdon**	manomètre de Bourdon	1.1.3-**8**

	ENGLISH	ESPAÑOL	FRANÇAIS
1.1.3-9	combined pressure and vacuum gauge *compound gauge*	**manovacuómetro** manómetro compuesto	manovacuomètre
1.1.3-10	compound gauge *combined pressure and vacuum gauge*	**manovacuómetro** manómetro compuesto	manovacuomètre
1.1.3-11	control pressure gauge	**medidor de presión para control automático**	manomètre à fonction(s) de commande
1.1.3-12	diaphragm	**diafragma**	membrane
1.1.3-13	diaphragm manometer *membrane manometer*	**manómetro de membrana**	manomètre à membrane
1.1.3-14	differential pressure	**presión diferencial**	pression différentielle
1.1.3-15	differential pressure gauge	**medidor de presión diferencial**	manomètre différentiel
1.1.3-16	duplex pressure gauge	**manómetro duplex**	manomètre duplex
1.1.3-17	dynamic head *velocity head*	**carga dinámica** altura dinámica	charge dynamique *hauteur dynamique*
1.1.3-18	dynamic pressure *velocity pressure*	**presión dinámica** presión debida a la velocidad	pression dynamique *pression due à la vitesse*
1.1.3-19	edgewise pressure gauge	**medidor de presión para encastrar** medidor de presión para panel	manomètre de profil
1.1.3-20	excess pressure	**sobrepresión**	surpression
1.1.3-21	gauge pressure	**presión manométrica** presión relativa	pression relative *pression effective*
1.1.3-22	hydrostatic pressure	**presión hidrostática**	pression hydrostatique
1.1.3-23	indicating pressure gauge	**medidor de presión** manómetro de aguja	manomètre indicateur
1.1.3-24	indicating pressure gauge with an elastic measuring element	**medidor de presión con elemento de medida elástico**	manomètre métallique indicateur
1.1.3-25	ionization vacuum gauge	**manómetro de ionización**	jauge à ionisation *manomètre à ionisation*
1.1.3-26	liquid-filled pressure gauge	**manómetro de glicerina**	manomètre à liquide amortisseur
1.1.3-27	liquid inlet pressure	**presión de entrada del líquido**	pression d'entrée du fluide
1.1.3-28	(liquid level) manometer *U-tube manometer*	**manómetro de columna de líquido**	manomètre à tube de liquide
1.1.3-29	liquid outlet pressure	**presión de salida del líquido**	pression de sortie du fluide frigoporteur
1.1.3-30	loss of head	**pérdida de carga**	perte de charge
1.1.3-31	membrane manometer *diaphragm manometer*	**manómetro de membrana**	manomètre à membrane
1.1.3-32	osmotic pressure	**presión osmótica**	pression osmotique
1.1.3-33	partial pressure	**presión parcial**	pression partielle
1.1.3-34	piezometer ring	**anillo piezométrico**	bague piézométrique
1.1.3-35	pointer	**indicador** puntero aguja indicadora	aiguille
1.1.3-36	pressure	**presión**	pression
1.1.3-37	pressure drop	**caída de presión**	chute de pression
1.1.3-38	pressure gauge	**manómetro**	manomètre
1.1.3-39	pressure loss	**pérdida de presión**	perte de pression
1.1.3-40	pressure-responsive element	**elemento sensible a la presión**	organe moteur

ENGLISH	ESPAÑOL	FRANÇAIS	
refrigerant inlet pressure	**presión de entrada del frigorígeno**	pression d'entrée du fluide frigorigène	1.1.3-**41**
relative vapour pressure	**presión de vapor relativa**	pression relative de vapeur	1.1.3-**42**
safety pattern gauge	**manómetro de seguridad**	manomètre de sureté	1.1.3-**43**
service gauge	**manómetro de montador**	manomètre de monteur	1.1.3-**44**
shank	**caja** cuerpo	raccord	1.1.3-**45**
stagnation pressure	**presión total** presión de estancamiento	pression d'arrêt	1.1.3-**46**
standard atmospheric pressure	**presión atmosférica normal**	pression atmosphérique normale	1.1.3-**47**
static head	**carga estática** altura estática	hauteur statique *charge statique*	1.1.3-**48**
static pressure	**presión estática**	pression statique	1.1.3-**49**
thermal conductivity vacuum gauge	**manómetro térmico**	jauge thermique *manomètre thermique*	1.1.3-**50**
total head	**carga total** altura total de columna de líquido altura total	charge totale	1.1.3-**51**
total pressure	**presión total**	pression totale	1.1.3-**52**
U-tube manometer *(liquid level) manometer*	**manómetro de columna de líquido**	manomètre à tube de liquide	1.1.3-**53**
ultra-high vacuum	**ultravacío**	ultravide	1.1.3-**54**
vacuum	**vacío** gas enrarecido	vide *gaz raréfié*	1.1.3-**55**
vacuum gauge	**manómetro de vacío**	jauge à vide *manomètre à vide*	1.1.3-**56**
vapour pressure	**presión de vapor**	pression de vapeur	1.1.3-**57**
velocity head *dynamic head*	**carga dinámica** altura dinámica	charge dynamique *hauteur dynamique*	1.1.3-**58**
velocity pressure *dynamic pressure*	**presión dinámica** presión debida a la velocidad	pression dynamique *pression due à la vitesse*	1.1.3-**59**
viscosity manometer	**manómetro de viscosidad**	manomètre à viscosité	1.1.3-**60**
wall (pressure) tapping	**toma de presión estática**	prise (de pression) à la paroi	1.1.3-**61**
water column	**columna de agua**	colonne d'eau	1.1.3-**62**

SECTION 1.1.4 *Humidity*	SUBCAPÍTULO 1.1.4 *Humedad*	SOUS-CHAPITRE 1.1.4 *Humidité*	
absolute humidity *water vapour concentration* ○	**humedad absoluta**	humidité absolue *concentration de la vapeur d'eau* ○	1.1.4-**1**
absorption hygrometer *chemical hygrometer*	**higrómetro de absorción**	hygromètre à absorption	1.1.4-**2**
air dry-bulb temperature *dry-bulb temperature*	**temperatura de bulbo seco** temperatura seca	température de bulbe sec *température sèche (de l'air)* ◑	1.1.4-**3**
air wet-bulb temperature *wet-bulb temperature*	**temperatura de bulbo húmedo** temperatura húmeda	température de bulbe humide *température humide (de l'air)* ◑	1.1.4-**4**
aspirated hygrometer *aspiration psychrometer*	**sicrómetro de aspiración**	psychromètre à aspiration *psychromètre ventilé*	1.1.4-**5**

	ENGLISH	ESPAÑOL	FRANÇAIS
1.1.4-6	aspiration psychrometer *aspirated hygrometer*	**sicrómetro de aspiración**	psychromètre à aspiration *psychromètre ventilé*
1.1.4-7	chemical hygrometer *absorption hygrometer*	**higrómetro de absorción**	hygromètre à absorption
1.1.4-8	degree of saturation *saturation ratio*	**grado de saturación**	degré de saturation
1.1.4-9	dehumidification	**deshumidificación**	déshumidification
1.1.4-10	dehumidify (to)	**deshumidificar**	déshumidifier
1.1.4-11	dehydrate (to)	**deshidratar**	déshydrater
1.1.4-12	dehydration *drying*	**deshidratación** secado	déshydratation *séchage*
1.1.4-13	desiccation	**desecación**	dessiccation
1.1.4-14	dew (1)	**rocío**	rosée
1.1.4-15	dew (2)	**rocío**	rosée
1.1.4-16	dew point *dew-point (temperature)*	**temperatura de punto de rocío** temperatura de rocío	température de rosée
1.1.4-17	dew-point depression	**separación del punto de rocío**	écart du point de rosée
1.1.4-18	dew-point hygrometer	**higrómetro de punto de rocío**	hygromètre à point de rosée
1.1.4-19	dew-point (temperature) *dew point*	**temperatura de punto de rocío** temperatura de rocío	température de rosée
1.1.4-20	dry air	**aire seco**	air sec
1.1.4-21	dry bulb	**bulbo seco**	bulbe sec
1.1.4-22	dry-bulb temperature *air dry-bulb temperature*	**temperatura de bulbo seco** temperatura seca	température de bulbe sec *température sèche (de l'air)* ○
1.1.4-23	dry-bulb thermometer	**termómetro de bulbo seco** termómetro seco	thermomètre (à bulbe) sec
1.1.4-24	drying *dehydration*	**deshidratación** secado	séchage *déshydratation*
1.1.4-25	electrical hygrometer	**higrómetro eléctrico**	hygromètre électrique
1.1.4-26	electrolytic hygrometer	**higrómetro electrolítico**	hygromètre électrolytique
1.1.4-27	fog *mist*	**niebla**	brouillard *buée*
1.1.4-28	frost point	**punto de escarcha**	point de gelée blanche *point de givre*
1.1.4-29	glaze	**capa de hielo** película de hielo	verglas *givre transparent*
1.1.4-30	hair hygrometer	**higrómetro de cabello**	hygromètre à cheveux
1.1.4-31	hoarfrost	**escarcha cristalina**	gelée blanche
1.1.4-32	humid *moist*	**húmedo**	humide
1.1.4-33	humid air	**aire húmedo**	air humide
1.1.4-34	humid volume ○	**volumen másico del aire húmedo** volumen específico del aire húmedo (referido al aire seco)	volume spécifique *volume massique rapporté à l'air sec*
1.1.4-35	humidification	**humectación** humidificación	humidification
1.1.4-36	humidify (to)	**humectar** humidificar	humidifier

ENGLISH	ESPAÑOL	FRANÇAIS	
humidity	**humedad**	humidité	1.1.4-**37**
humidity ratio *mixing ratio* *moisture content (1)*	**humedad específica** fracción másica de humedad	humidité spécifique (air humide)	1.1.4-**38**
hygrometer	**higrómetro**	hygromètre	1.1.4-**39**
hygrometry	**higrometría**	hygrométrie	1.1.4-**40**
hygroscopic	**higroscópico**	hygroscopique	1.1.4-**41**
mist *fog*	**niebla**	buée *brouillard*	1.1.4-**42**
mixing ratio *humidity ratio* *moisture content (1)*	**humedad específica** fracción másica de humedad	humidité spécifique (air humide)	1.1.4-**43**
moist *humid*	**húmedo**	humide	1.1.4-**44**
moisture carry-over	**arrastre de gotitas de agua**	entraînement de gouttelettes (d'eau)	1.1.4-**45**
moisture content (1) *humidity ratio* *mixing ratio*	**humedad específica** fracción másica de humedad	humidité spécifique (air humide)	1.1.4-**46**
moisture content (2) *water content*	**contenido de agua** contenido de humedad	teneur en eau (substance solide)	1.1.4-**47**
moisture indicator	**indicador de humedad**	indicateur d'humidité	1.1.4-**48**
moisture transfer	**transmisión de humedad** transporte de humedad	transfert d'humidité *transport d'humidité*	1.1.4-**49**
mole fraction of the water vapour	**fracción molar de vapor de agua** título molar de vapor de agua	fraction molaire de vapeur d'eau *titre molaire de vapeur d'eau* ○	1.1.4-**50**
organic hygrometer	**higrómetro orgánico**	hygromètre organique	1.1.4-**51**
psychrometer	**sicrómetro**	psychromètre	1.1.4-**52**
psychrometric chart	**diagrama sicrométrico**	diagramme psychrométrique	1.1.4-**53**
psychrometrics *psychrometry*	**sicrometría**	psychrométrie	1.1.4-**54**
psychrometry *psychrometrics*	**sicrometría**	psychrométrie	1.1.4-**55**
relative humidity	**grado higrométrico** humedad relativa	humidité relative	1.1.4-**56**
relative humidity with respect to ice (W.M.O.)	**humedad relativa con relación al hielo** **(OMM)**	humidité relative par rapport à la glace (O.M.M.)	1.1.4-**57**
relative humidity with respect to water (W.M.O.) ○	**humedad relativa con relación al agua** **(OMM)**	humidité relative par rapport à l'eau (O.M.M.) ○	1.1.4-**58**
rime	**escarcha opaca**	givre blanc	1.1.4-**59**
saturated air ○ *saturated humid air*	**aire húmedo saturado** aire saturado	air humide saturé *air saturé* ○	1.1.4-**60**
saturated humid air *saturated air* ○	**aire húmedo saturado** aire saturado	air humide saturé *air saturé* ○	1.1.4-**61**
saturation ratio *degree of saturation*	**grado de saturación**	degré de saturation	1.1.4-**62**
sling hygrometer *sling psychrometer*	**higrómetro de carraca** sicrómetro de carraca	psychromètre à rotation *psychromètre fronde*	1.1.4-**63**
sling psychrometer *sling hygrometer*	**higrómetro de carraca** sicrómetro de carraca	psychromètre à rotation *psychromètre fronde*	1.1.4-**64**

	ENGLISH	ESPAÑOL	FRANÇAIS
1.1.4-**65**	specific humidity *water vapour content* ◐	**humedad específica** contenido de vapor de agua	humidité spécifique *teneur en vapeur d'eau* ◐
1.1.4-**66**	steam *water vapour*	**vapor de agua**	vapeur d'eau
1.1.4-**67**	supersaturated air	**aire sobresaturado**	air sursaturé
1.1.4-**68**	sweating	**condensación de agua (sobre una superficie)**	condensation d'eau (sur une surface)
1.1.4-**69**	water content *moisture content (2)*	**contenido de agua** contenido de humedad	teneur en eau (substance solide)
1.1.4-**70**	water vapour *steam*	**vapor de agua**	vapeur d'eau
1.1.4-**71**	water vapour concentration ○ *absolute humidity*	**humedad absoluta**	humidité absolue *concentration de la vapeur d'eau* ○
1.1.4-**72**	water vapour content ◐ *specific humidity*	**humedad específica** contenido de vapor de agua	humidité spécifique *teneur en vapeur d'eau* ◐
1.1.4-**73**	wet bulb	**bulbo húmedo**	bulbe humide
1.1.4-**74**	wet-bulb depression	**diferencia sicrométrica**	différence psychrométrique
1.1.4-**75**	wet-bulb temperature *air wet-bulb temperature*	**temperatura de bulbo húmedo** temperatura húmeda	température de bulbe humide *température humide (de l'air)* ◐
1.1.4-**76**	wet-bulb thermometer	**termómetro de bulbo húmedo** termómetro húmedo	thermomètre (à bulbe) humide

	SECTION 1.1.5 *Heat*	**SUBCAPÍTULO 1.1.5** *Calor*	**SOUS-CHAPITRE 1.1.5** *Chaleur*
1.1.5-**1**	amount of heat *heat quantity*	**cantidad de calor**	quantité de chaleur
1.1.5-**2**	area coefficient of heat loss	**coeficiente superficial de pérdida de calor**	coefficient surfacique de déperdition thermique
1.1.5-**3**	bolometer	**bolómetro**	bolomètre
1.1.5-**4**	calorimeter	**calorímetro**	calorimètre
1.1.5-**5**	calorimetry	**calorimetria**	calorimétrie
1.1.5-**6**	coefficient of expansion	**coeficiente de dilatación**	coefficient de dilatation
1.1.5-**7**	density of heat flow rate *heat flux*	**densidad de flujo térmico**	densité de flux thermique *flux thermique surfacique*
1.1.5-**8**	exergetic efficiency *second-law efficiency*	**rendimiento exergético**	rendement exergétique
1.1.5-**9**	heat	**calor**	chaleur
1.1.5-**10**	heat exchanger effectiveness *thermal efficiency (1)*	**rendimiento térmico**	efficacité d'un échangeur thermique
1.1.5-**11**	heat flux *density of heat flow rate*	**densidad de flujo térmico**	densité de flux thermique *flux thermique surfacique*
1.1.5-**12**	heat gain *heat uptake* ◐	**aporte de calor** entradas de calor ganancia de calor	apport de chaleur *entrée de chaleur* *gain de chaleur*
1.1.5-**13**	heat loss	**pérdida de calor**	perte de chaleur
1.1.5-**14**	heat quantity *amount of heat*	**cantidad de calor**	quantité de chaleur

ENGLISH	ESPAÑOL	FRANÇAIS	
heat storage *thermal storage*	**almacenamiento térmico** acumulación de calor	accumulation de chaleur *stockage thermique*	1.1.5-**15**
heat uptake ○ *heat gain*	**aporte de calor** entradas de calor ganancia de calor	apport de chaleur *entrée de chaleur* *gain de chaleur*	1.1.5-**16**
heating	**calefacción**	chauffage *échauffement*	1.1.5-**17**
linear thermal resistance	**resistencia térmica lineal**	résistance thermique linéique	1.1.5-**18**
linear thermal transmittance	**transmitancia térmica lineal**	coefficient linéique de transmission thermique	1.1.5-**19**
overall heat transfer coefficient *thermal transmittance*	**transmitancia térmica** coeficiente global de transmisión de calor coeficiente global de transmisión térmica	coefficient global de transfert de chaleur *coefficient global d'échange thermique* *transmittance thermique* ○	1.1.5-**20**
reheat (to)	**recalentar**	réchauffer	1.1.5-**21**
reject heat	**calor desprendido** calor rechazado	rejet thermique	1.1.5-**22**
room calorimeter	**cámara calorimétrica**	chambre calorimétrique	1.1.5-**23**
second-law efficiency *exergetic efficiency*	**rendimiento exergético**	rendement exergétique	1.1.5-**24**
sensible heat	**calor sensible**	chaleur sensible	1.1.5-**25**
thermal conductance	**conductancia térmica**	conductance thermique	1.1.5-**26**
thermal diffusivity	**difusividad térmica** difusión térmica	diffusivité thermique	1.1.5-**27**
thermal efficiency (1) *heat exchanger effectiveness*	**rendimiento térmico**	efficacité d'un échangeur thermique	1.1.5-**28**
thermal efficiency (2)	**rendimiento térmico**	rendement thermodynamique	1.1.5-**29**
thermal effusivity	**efusividad térmica**	effusivité thermique	1.1.5-**30**
(thermal) expansion	**dilatación (térmica)**	dilatation (thermique)	1.1.5-**31**
thermal resistance	**resistencia térmica** coeficiente de resistencia térmica	résistance thermique	1.1.5-**32**
thermal storage *heat storage*	**almacenamiento térmico** acumulación de calor	accumulation de chaleur *stockage thermique*	1.1.5-**33**
thermal transmittance *overall heat transfer coefficient*	**transmitancia térmica** coeficiente global de transmisión de calor coeficiente global de transmisión térmica	coefficient global de transfert de chaleur *coefficient global d'échange thermique* *transmittance thermique* ○	1.1.5-**34**
volume coefficient of heat loss	**coeficiente volumétrico de pérdida de calor**	coefficient volumique de déperdition thermique	1.1.5-**35**
waste heat	**calor cedido** calor perdido	chaleur perdue	1.1.5-**36**

SECTION 1.1.6 *Fluid flow*	SUBCAPÍTULO 1.1.6 *Flujo de fluidos*	SOUS-CHAPITRE 1.1.6 *Ecoulement des fluides*	
absolute viscosity ○ *dynamic viscosity* *coefficient of viscosity* ○	**viscosidad dinámica**	viscosité dynamique *coefficient de viscosité* ○	1.1.6-**1**
airflow	**caudal de aire**	débit d'air	1.1.6-**2**
anemometer	**anemómetro**	anémomètre	1.1.6-**3**

ENGLISH	ESPAÑOL	FRANÇAIS
1.1.6-**4** annular flow	**flujo anular**	écoulement annulaire
1.1.6-**5** aspiration	**aspiración** succión	aspiration *succion*
1.1.6-**6** balanced flow	**caudal equilibrado** flujo equilibrado	flux équilibré
1.1.6-**7** boundary layer	**capa límite**	couche limite
1.1.6-**8** bubble flow	**flujo con burbujas** corriente con burbujas	écoulement à bulles
1.1.6-**9** buoyancy	**flotabilidad**	flottabilité
1.1.6-**10** carrier ring	**arandela de soporte**	bague porteuse
1.1.6-**11** coefficient of viscosity ○ *dynamic viscosity* *absolute viscosity* ○	**viscosidad dinámica**	viscosité dynamique *coefficient de viscosité* ○
1.1.6-**12** concentric orifice plate	**placa con orificio concéntrico** placa de orificio concéntrico	diaphragme concentrique
1.1.6-**13** Couette flow	**flujo Couette** flujo de Couette	écoulement de Couette
1.1.6-**14** counterflow	**contracorriente**	contre-courant
1.1.6-**15** critical velocity	**velocidad crítica**	vitesse critique
1.1.6-**16** cross flow	**corriente cruzada** flujo cruzado	écoulements croisés *écoulements transversaux*
1.1.6-**17** cup anemometer	**anemoscopio** anemómetro de cazoletas	anémomètre à coupelles
1.1.6-**18** deflecting vane anemometer	**anemómetro de paleta**	anémomètre à palette
1.1.6-**19** differential pressure device	**sonda deprimógena** elemento deprimógeno	appareil déprimogène
1.1.6-**20** digital anemometer	**anemómetro digital**	anémomètre à impulsions
1.1.6-**21** dispersed flow	**corriente dispersa** flujo disperso	écoulement dispersé
1.1.6-**22** dynamic loss	**pérdida dinámica**	perte dynamique
1.1.6-**23** dynamic viscosity *absolute viscosity* ○ *coefficient of viscosity* ○	**viscosidad dinámica**	viscosité dynamique *coefficient de viscosité* ○
1.1.6-**24** eddy flow ○ *turbulent flow*	**flujo turbulento** corriente turbulenta	écoulement turbulent
1.1.6-**25** electromagnetic flowmeter	**caudalímetro electromagnético**	débitmètre électromagnétique
1.1.6-**26** electronic anemometer	**anemómetro electrónico**	anémomètre électronique
1.1.6-**27** emulsion flow	**flujo con emulsión**	écoulement à émulsion *écoulement mousseux* ○
1.1.6-**28** face tube ○ *Pitot tube*	**tubo de Pitot**	tube de Pitot
1.1.6-**29** flow conditioner *flow straightener*	**acondicionador de flujo** enderezador de flujo	tranquilliseur de débit
1.1.6-**30** flow nozzle	**tobera**	tuyère *ajutage* ○
1.1.6-**31** flow pattern	**tipo de flujo** clase de flujo	configuration d'écoulement
1.1.6-**32** flow profile	**perfil de flujo** perfil de velocidades	profil des vitesses

ENGLISH	ESPAÑOL	FRANÇAIS	
flow rate (of a fluid) *rate of flow*	**caudal (de un fluido)** gasto	débit	1.1.6-**33**
flow stabilizer	**estabilizador de flujo**	stabilisateur de débit	1.1.6-**34**
flow straightener *flow conditioner*	**acondicionador de flujo** enderezador de flujo	tranquilliseur de débit	1.1.6-**35**
flowmeter	**caudalímetro** fluxómetro	débitmètre	1.1.6-**36**
fluid flow	**flujo de fluidos** corriente derrame	écoulement d'un fluide	1.1.6-**37**
fluidized bed	**lecho fluidizado**	lit fluidisé	1.1.6-**38**
friction factor	**coeficiente de rozamiento**	coefficient de frottement	1.1.6-**39**
friction loss	**pérdida de carga por rozamiento**	perte de charge frictionnelle *perte de charge par frottement*	1.1.6-**40**
frictional resistance	**resistencia de rozamiento**	résistance de frottement	1.1.6-**41**
fully developed velocity distribution	**distribución de velocidades estabilizada**	répartition pleinement développée (ou pleinement établie) de la vitesse	1.1.6-**42**
heated thermometer anemometer ○ *thermal anemometer* *hot-wire anemometer*	**anemómetro térmico** anemómetro de cuerpo caliente anemómetro de hilo caliente	anémomètre thermique *anémomètre à corps chaud* *anémomètre à fil chaud*	1.1.6-**43**
hot-wire anemometer *thermal anemometer* *heated thermometer anemometer* ○	**anemómetro térmico** anemómetro de cuerpo caliente anemómetro de hilo caliente	anémomètre thermique *anémomètre à corps chaud* *anémomètre à fil chaud*	1.1.6-**44**
hydraulic diameter	**diámetro hidráulico**	diamètre hydraulique	1.1.6-**45**
interfacial tension ○ *surface tension*	**tensión superficial** tension interfacial	tension superficielle *tension interfaciale* ○	1.1.6-**46**
kinematic viscosity	**viscosidad cinemática**	viscosité cinématique	1.1.6-**47**
lack of miscibility	**falta de miscibilidad**	lacune de miscibilité *démixtion* ○	1.1.6-**48**
laminar flow	**corriente laminar** flujo laminar	écoulement laminaire	1.1.6-**49**
mass flow rate	**caudal másico**	débit masse	1.1.6-**50**
mass flux *mass velocity*	**velocidad másica**	vitesse massique *densité de flux massique*	1.1.6-**51**
mass velocity *mass flux*	**velocidad másica**	vitesse massique *densité de flux massique*	1.1.6-**52**
mean flow rate	**caudal medio**	débit moyen	1.1.6-**53**
mechanical anemometer	**anemómetro mecánico**	anémomètre mécanique	1.1.6-**54**
meter tube	**tubo patrón**	tube de mesurage	1.1.6-**55**
miscibility	**miscibilidad**	miscibilité	1.1.6-**56**
mist flow	**flujo de niebla**	(écoulement de) buée (écoulement de) brouillard *écoulement vésiculaire*	1.1.6-**57**
mixture	**mezcla**	mélange	1.1.6-**58**
molecular flow	**corriente molecular** flujo molecular	écoulement (en régime) moléculaire *flux moléculaire* ○	1.1.6-**59**
multi-phase flow	**flujo multifásico**	écoulement multiphasique	1.1.6-**60**
nozzle	**tobera**	tuyère	1.1.6-**61**

	ENGLISH	ESPAÑOL	FRANÇAIS
1.1.6-62	orifice plate	**orificio calibrado**	diaphragme de mesure *orifice de jaugeage* ○
1.1.6-63	parallel flow	**flujo a equicorriente**	écoulements parallèles et de même sens *cocourant* *équicourant*
1.1.6-64	Pitot tube *face tube* ○	**tubo de Pitot**	tube de Pitot
1.1.6-65	plug flow	**corriente con taponamientos alternativos** flujo con taponamientos alternativos	écoulement à bouchons
1.1.6-66	porous plug	**tapón poroso**	bouchon poreux
1.1.6-67	pulsating flow	**corriente pulsátil** flujo pulsátil	écoulement pulsatoire
1.1.6-68	pulsating flow of mean constant flow rate	**flujo pulsante de caudal medio constante**	écoulement pulsatoire à débit moyen constant
1.1.6-69	rate of flow *flow rate (of a fluid)*	**caudal (de un fluido)** gasto	débit
1.1.6-70	regular velocity distribution	**distribución de velocidad regular** distribución de velocidad uniforme	répartition régulière de la vitesse
1.1.6-71	revolving vane anemometer	**anemómetro de molinete**	anémomètre à hélice
1.1.6-72	Reynolds number (Re)	**número de Reynolds**	nombre de Reynolds (Re)
1.1.6-73	roughness factor	**coeficiente de rugosidad**	rugosité relative
1.1.6-74	slug flow	**flujo bifásico discontínuo**	***
1.1.6-75	sonic velocity *speed of sound*	**velocidad del sonido**	célérité du son *vitesse du son* ○
1.1.6-76	speed of sound *sonic velocity*	**velocidad del sonido**	célérité du son *vitesse du son* ○
1.1.6-77	stagnation point	**punto de estancamiento**	point d'arrêt
1.1.6-78	stagnation temperature *total temperature* ○	**temperatura total**	température d'arrêt *température totale* ○
1.1.6-79	steady flow	**flujo estacionario**	écoulement permanent
1.1.6-80	straight length	**tramo recto**	longueur droite
1.1.6-81	stratified flow	**corriente estratificada** flujo estratificado	écoulement stratifié
1.1.6-82	subsonic flow	**flujo subsónico**	écoulement subsonique
1.1.6-83	supersonic flow	**flujo supersónico**	écoulement supersonique
1.1.6-84	surface tension *interfacial tension* ○	**tensión superficial** tension interfacial	tension superficielle *tension interfaciale* ○
1.1.6-85	thermal anemometer *hot-wire anemometer* *heated thermometer anemometer* ○	**anemómetro térmico** anemómetro de cuerpo caliente anemómetro de hilo caliente	anémomètre thermique *anémomètre à corps chaud* *anémomètre à fil chaud*
1.1.6-86	thin orifice plate	**placa con orificio delgado** placa de orificio delgado	diaphragme en mince paroi
1.1.6-87	throttling	**estrangulación** estrechamiento	étranglement
1.1.6-88	total temperature ○ *stagnation temperature*	**temperatura total**	température d'arrêt *température totale* ○
1.1.6-89	transition flow	**flujo transitorio**	écoulement de transition
1.1.6-90	turbine flowmeter	**caudalímetro de turbina**	débitmètre à hélice *débitmètre à turbine*

ENGLISH	ESPAÑOL	FRANÇAIS	
turbulent flow *eddy flow* ◐	**flujo turbulento** corriente turbulenta	écoulement turbulent	1.1.6-**91**
two-phase flow	**corriente bifásica** flujo bifásico	écoulement diphasique	1.1.6-**92**
ultrasonic flowmeter	**caudalímetro ultrasónico**	débitmètre à ultrasons	1.1.6-**93**
universal head loss coefficient	**coeficiente universal de pérdida de presión**	coefficient universel de perte de charge	1.1.6-**94**
unsteady flow	**flujo no estacionario**	écoulement non permanent	1.1.6-**95**
vane anemometer	**anemómetro de molinete**	anémomètre à moulinet	1.1.6-**96**
velocity distribution	**distribución de velocidades**	répartition des vitesses	1.1.6-**97**
velocity of flow	**velocidad de corriente** velocidad de flujo	vitesse d'écoulement	1.1.6-**98**
velocity profile	**perfil de velocidades**	profil des vitesses	1.1.6-**99**
Venturi tube	**tubo de Venturi** venturi	tube de Venturi	1.1.6-**100**
viscometer *viscosimeter*	**viscosímetro**	viscosimètre	1.1.6-**101**
viscosimeter *viscometer*	**viscosímetro**	viscosimètre	1.1.6-**102**
viscosity	**viscosidad**	viscosité	1.1.6-**103**
viscous flow	**corriente viscosa** flujo viscoso	écoulement visqueux	1.1.6-**104**
volume flow rate *volumetric flow rate*	**caudal volumétrico**	débit volume	1.1.6-**105**
volumetric flow rate *volume flow rate*	**caudal volumétrico**	débit volume	1.1.6-**106**
vortex flowmeter	**caudalímetro de vórtice**	débitmètre à vortex	1.1.6-**107**
wavy flow	**corriente ondulada** flujo ondulado	écoulement ondulé	1.1.6-**108**

SECTION 1.1.7 *Metrology – Measuring apparatus*	SUBCAPÍTULO 1.1.7 *Metrología – Aparatos de medida*	SOUS-CHAPITRE 1.1.7 *Métrologie – Appareils de mesure*	
accuracy	**precisión** exactitud	exactitude	1.1.7-**1**
accuracy of measurement	**precisión de la medida** exactitud de la medida	exactitude de mesure	1.1.7-**2**
adjustment (of a measuring instrument)	**ajuste (de un instrumento de medida)**	ajustage (d'un instrument de mesure)	1.1.7-**3**
analogue indicating instrument *analogue measuring instrument*	**instrumento de medida analógico**	appareil de mesure (à affichage) analogique	1.1.7-**4**
analogue measuring instrument *analogue indicating instrument*	**instrumento de medida analógico**	appareil de mesure (à affichage) analogique	1.1.7-**5**
base quantity	**magnitud fundamental**	grandeur de base	1.1.7-**6**
calibration	**calibración**	étalonnage	1.1.7-**7**
case	**caja**	boîtier	1.1.7-**8**
certified standard instrument	**instrumento con certificado standard**	instrument étalonné agréé	1.1.7-**9**
chart	**dispositivo de almacenamiento** registro	support d'enregistrement	1.1.7-**10**

	ENGLISH	ESPAÑOL	FRANÇAIS
1.1.7-**11**	controlled variable	**magnitud regulada**	grandeur réglée
1.1.7-**12**	correcting variable	**magnitud reguladora**	grandeur réglante
1.1.7-**13**	derived quantity	**magnitud derivada**	grandeur dérivée
1.1.7-**14**	detecting element *sensor* *sensing element* ◑	**sensor** detector elemento sensible	capteur *élément sensible*
1.1.7-**15**	detector	**detector**	détecteur
1.1.7-**16**	dial	**dial**	cadran
1.1.7-**17**	digital indicating instrument *digital measuring instrument*	**instrumento de medida digital**	appareil de mesure (à affichage) numérique
1.1.7-**18**	digital measuring instrument *digital indicating instrument*	**instrumento de medida digital**	appareil de mesure (à affichage) numérique
1.1.7-**19**	dimensionless quantity *quantity of dimension one* ○	**magnitud adimensional**	grandeur adimensionnelle *grandeur de dimension un*
1.1.7-**20**	displaying device *indicating device*	**indicador** display indicador numérico lector pantalla	dispositif d'affichage *dispositif indicateur*
1.1.7-**21**	displaying (measuring) instrument *indicating (measuring) instrument*	**lector** indicador indicador numérico pantalla	appareil (de mesure) afficheur *appareil (de mesure) indicateur*
1.1.7-**22**	error of measurement	**error de la medida**	erreur de mesure *erreur de mesurage*
1.1.7-**23**	flange	**brida**	collerette
1.1.7-**24**	gauging (of a measuring instrument)	**graduación** aforo	calibrage (d'un instrument de mesure)
1.1.7-**25**	indicating device *displaying device*	**indicador** display indicador numérico lector pantalla	dispositif d'affichage *dispositif indicateur*
1.1.7-**26**	indicating (measuring) instrument *displaying (measuring) instrument*	**lector** indicador indicador numérico pantalla	appareil (de mesure) afficheur *appareil (de mesure) indicateur*
1.1.7-**27**	indication (of a measuring instrument)	**lectura (de un instrumento de medida)**	indication (d'un instrument de mesure)
1.1.7-**28**	influence quantity	**magnitud de influencia** magnitud influyente	grandeur d'influence
1.1.7-**29**	input (variable)	**entrada (variable)**	grandeur d'entrée
1.1.7-**30**	integrating (measuring) instrument	**integrador**	appareil (de mesure) intégrateur
1.1.7-**31**	International System of Units (SI)	**Sistema Internacional de Unidades (SI)**	Système International d'unités (SI)
1.1.7-**32**	intrinsic error (of a measuring instrument)	**error intrínseco**	erreur intrinsèque (d'un instrument de mesure)
1.1.7-**33**	limiting conditions	**condiciones límite**	conditions limites
1.1.7-**34**	(measurable) quantity	**magnitud (medible)**	grandeur (mesurable)
1.1.7-**35**	measurand	**cantidad**	mesurande
1.1.7-**36**	measurement	**medida**	mesurage
1.1.7-**37**	measurement procedure	**procedimiento de medida**	mode opératoire (du mesurage)

ENGLISH	ESPAÑOL	FRANÇAIS	
measurement value (of a quantity) *result of a measurement*	**medición** medida resultado de una medida	mesure *résultat du mesurage*	1.1.7-**38**
measuring instrument	**instrumento de medida**	instrument de mesure *appareil de mesure*	1.1.7-**39**
measuring range *working range*	**rango de medida**	étendue de mesure *plage de mesure*	1.1.7-**40**
measuring transducer	**transductor**	transducteur de mesurage *transducteur de mesure* *jauge de mesurage*	1.1.7-**41**
method of measurement	**método de medida**	méthode de mesure	1.1.7-**42**
metrology	**metrología**	métrologie	1.1.7-**43**
nominal range	**rango nomimal**	calibre *plage*	1.1.7-**44**
output (variable)	**magnitud de salida**	grandeur de sortie	1.1.7-**45**
precision	**precisión**	précision	1.1.7-**46**
pressure sensor	**medidor de presión**	capteur de pression *jauge de pression*	1.1.7-**47**
quantity of dimension one ○ *dimensionless quantity*	**magnitud adimensional**	grandeur adimensionnelle *grandeur de dimension un*	1.1.7-**48**
random error	**error aleatorio**	erreur aléatoire	1.1.7-**49**
range of indication	**rango de lectura** rango de las indicaciones	étendue des indications *étendue d'échelle*	1.1.7-**50**
rated operating conditions	**condiciones de funcionamiento especificadas** condiciones asignadas de funcionamiento	conditions assignées de fonctionnement	1.1.7-**51**
recording device	**registrador** grabador	dispositif enregistreur	1.1.7-**52**
recording duration	**duración de registro**	durée d'enregistrement	1.1.7-**53**
recording interval	**intervalo de grabación** lapso de grabación	période d'enregistrement *intervalle d'enregistrement*	1.1.7-**54**
recording (measuring) instrument	**registrador** grabador instrumento de almacenado	appareil (de mesure) enregistreur	1.1.7-**55**
relative error	**error relativo**	erreur relative	1.1.7-**56**
repeatability of measurements	**repetibilidad de las medidas**	répétabilité des mesurages	1.1.7-**57**
reproducibility of measurements	**reproducibilidad de las medidas**	reproductibilité des mesurages	1.1.7-**58**
resolution	**resolución**	résolution	1.1.7-**59**
response time	**tiempo de respuesta**	temps de réponse	1.1.7-**60**
result of a measurement *measurement value (of a quantity)*	**medición** medida resultado de una medida	résultat du mesurage *mesure*	1.1.7-**61**
scale division	**división de la escala**	division	1.1.7-**62**
scale length	**longitud de la escala**	longueur d'échelle	1.1.7-**63**
scale (of a measuring instrument)	**escala (de un instrumento de medida)**	échelle (d'un instrument de mesure)	1.1.7-**64**
sensitivity	**sensibilidad**	sensibilité	1.1.7-**65**
sensing element ○ *sensor* *detecting element*	**sensor** detector elemento sensible	capteur *élément sensible*	1.1.7-**66**

	ENGLISH	ESPAÑOL	FRANÇAIS
1.1.7-**67**	sensor *detecting element* *sensing element* ◐	**sensor** detector elemento sensible	capteur *élément sensible*
1.1.7-**68**	span	**intervalo de medida**	intervalle de mesure
1.1.7-**69**	stability	**estabilidad**	constance
1.1.7-**70**	storage and transport conditions	**condiciones de almacenamiento y transporte**	conditions de stockage et de transport
1.1.7-**71**	symbol of unit (of measurement)	**símbolo de una unidad**	symbole d'une unité (de mesure)
1.1.7-**72**	temperature sensor	**sensor de temperatura**	capteur de température
1.1.7-**73**	transducer	**transductor (de medida)**	transducteur
1.1.7-**74**	uncertainty of measurement	**incertidumbre de la medida**	incertitude de mesure
1.1.7-**75**	unit (of measurement)	**unidad (de medida)**	unité (de mesure)
1.1.7-**76**	user adjustment (of a measuring instrument)	**ajuste hecho por el usuario (de un instrumento de medida)**	réglage (d'un instrument de mesure)
1.1.7-**77**	value (of a quantity)	**cantidad (de una magnitud)**	valeur (d'une grandeur)
1.1.7-**78**	working range *measuring range*	**rango de medida**	étendue de mesure *plage de mesure*

SECTION 1.2 *Thermodynamic properties and processes* **SECTION 1.2.1** *Thermodynamic properties and processes: general background*	**SUBCAPÍTULO 1.2** *Propiedades y transformaciones termodinámicas* **SUBCAPÍTULO 1.2.1** *Generalidades sobre propiedades termodinámicas y procesos*	**SOUS-CHAPITRE 1.2** *Propriétés et transformations thermodynamiques* **SOUS-CHAPITRE 1.2.1** *Généralités sur les propriétés et les transformations thermodynamiques*

	ENGLISH	ESPAÑOL	FRANÇAIS
1.2.1-**1**	adiabatic	**adiabático**	adiabatique
1.2.1-**2**	availability *exergy*	**exergía** energía utilizable	exergie
1.2.1-**3**	availability destruction	**pérdida de exergía**	perte d'exergie *destruction d'exergie*
1.2.1-**4**	boundary conditions	**condiciones límites**	conditions aux limites
1.2.1-**5**	closed process	**transformación cerrada**	système fermé
1.2.1-**6**	coefficient of compressibility ◐ *compressibility factor*	**coeficiente de compresibilidad**	facteur de compressibilité
1.2.1-**7**	cold reservoir *low-temperature reservoir* *heat source (2)* ◐	**fuente fría** foco frío (en el sentido termodinámico) fuente de frío	source froide (absorbant la chaleur)
1.2.1-**8**	cold source ○ *heat sink*	**sumidero de calor** foco frío fuente de frío	puits de chaleur *source de froid (au sens banal)* ◐
1.2.1-**9**	compressibility factor *coefficient of compressibility* ◐	**coeficiente de compresibilidad**	facteur de compressibilité
1.2.1-**10**	corresponding state(s)	**estados correspondientes**	états correspondants
1.2.1-**11**	cyclic thermal conditions ◐	**ciclo** régimen cíclico	régime cyclique *régime variable*
1.2.1-**12**	energy breakdown (in exergetic sense) ◐	**degradación de la energía**	dégradation de l'énergie (dans le sens d'exergie)

20

ENGLISH	ESPAÑOL	FRANÇAIS	
energy conservation (1)	**conservación de la energía**	conservation de l'énergie	1.2.1-**13**
energy conservation (2) *first law (of thermodynamics)* *equivalence principle* ● *first principle (of thermodynamics)* ●	**primer principio (de la termodinámica)** principio de equivalencia	premier principe (de la thermodynamique) *principe de l'équivalence* ●	1.2.1-**14**
energy level *energy state*	**nivel de la energía**	niveau d'énergie	1.2.1-**15**
energy state *energy level*	**nivel de la energía**	niveau d'énergie	1.2.1-**16**
enthalpy	**entalpía**	enthalpie	1.2.1-**17**
entropy	**entropía**	entropie	1.2.1-**18**
entropy generation *entropy production*	**aumento de entropía** incremento de entropía	création d'entropie	1.2.1-**19**
entropy production *entropy generation*	**aumento de entropía** incremento de entropía	création d'entropie	1.2.1-**20**
equivalence principle ● *first law (of thermodynamics)* *energy conservation (2)* *first principle (of thermodynamics)* ●	**primer principio (de la termodinámica)** principio de equivalencia	premier principe (de la thermodynamique) *principe de l'équivalence* ●	1.2.1-**21**
exergetic efficiency *second-law efficiency*	**rendimiento exergético**	rendement exergétique	1.2.1-**22**
exergy *availability*	**exergía** energía utilizable	exergie	1.2.1-**23**
first law (of thermodynamics) *energy conservation (2)* *equivalence principle* ● *first principle (of thermodynamics)* ●	**primer principio (de la termodinámica)** principio de equivalencia	premier principe (de la thermodynamique) *principe de l'équivalence* ●	1.2.1-**24**
first principle (of thermodynamics) ● *first law (of thermodynamics)* *energy conservation (2)* *equivalence principle* ●	**primer principio (de la termodinámica)** principio de equivalencia	premier principe (de la thermodynamique) *principe de l'équivalence* ●	1.2.1-**25**
fugacity	**fugacidad**	fugacité	1.2.1-**26**
heat equivalent of work ○	**equivalente térmico del trabajo**	équivalent calorifique de travail ○	1.2.1-**27**
heat reservoir (1)	**fuente de calor** foco de calor (en el sentido termo- dinámico)	source de chaleur (au sens thermo- dynamique)	1.2.1-**28**
heat reservoir (2) *high-temperature reservoir* *hot reservoir*	**foco caliente** fuente caliente (en el sentido termo- dinámica)	source chaude	1.2.1-**29**
heat sink *cold source* ○	**sumidero de calor** foco frío fuente de frío	puits de chaleur *source de froid (au sens banal)* ●	1.2.1-**30**
heat source (1)	**foco caliente** fuente caliente (en el sentido termo- dinámica)	source de chaleur (au sens banal)	1.2.1-**31**
heat source (2) *cold reservoir* *low-temperature reservoir*	**fuente fría** foco frío (en el sentido termodinámico) fuente de frío	source froide (absorbant la chaleur)	1.2.1-**32**
high-temperature reservoir *heat reservoir (2)* *hot reservoir*	**foco caliente** fuente caliente (en el sentido termo- dinámica)	source chaude	1.2.1-**33**
hot reservoir *heat reservoir (2)* *high-temperature reservoir*	**foco caliente** fuente caliente (en el sentido termo- dinámica)	source chaude	1.2.1-**34**

	ENGLISH	ESPAÑOL	FRANÇAIS
1.2.1-35	ideal gas	**gas perfecto**	gaz idéal
1.2.1-36	(ideal) gas constant *(perfect) gas constant* ○	**constante de los gases perfectos**	constante des gaz idéaux (ou idéals) *constante des gaz parfaits*
1.2.1-37	(ideal) gas equation *(perfect) gas equation* ○	**ecuación de los gases perfectos**	équation des gaz idéaux (ou idéals) *équation des gaz parfaits*
1.2.1-38	ideal multistage compression	**compresión multietapa ideal**	compression polyétagée idéale
1.2.1-39	internal energy *intrinsic energy* ○	**energía interna**	énergie interne
1.2.1-40	intrinsic energy ○ *internal energy*	**energía interna**	énergie interne
1.2.1-41	irreversibility	**irreversibilidad**	irréversibilité
1.2.1-42	irreversible process	**transformación irreversible**	transformation irréversible
1.2.1-43	isolated system	**sistema aislado**	système isolé
1.2.1-44	law of corresponding states	**ley de los estados correspondientes**	loi des états correspondants
1.2.1-45	low-temperature reservoir *cold reservoir* *heat source (2)* ○	**fuente fría** foco frío (en el sentido termodinámico) fuente de frío	source froide (absorbant la chaleur)
1.2.1-46	mechanical equivalent of heat ○	**equivalente mecánico del calor** equivalente mecánico del trabajo	équivalent mécanique de chaleur ○
1.2.1-47	nonequilibrium thermodynamics	**termodinámica de procesos irreversibles**	thermodynamique des processus irréversibles
1.2.1-48	open process	**transformación abierta**	système ouvert
1.2.1-49	perfect gas	**gas perfecto**	gaz parfait
1.2.1-50	(perfect) gas constant ○ *(ideal) gas constant*	**constante de los gases perfectos**	constante des gaz idéaux (ou idéals) *constante des gaz parfaits*
1.2.1-51	(perfect) gas equation ○ *(ideal) gas equation*	**ecuación de los gases perfectos**	équation des gaz idéaux (ou idéals) *équation des gaz parfaits*
1.2.1-52	permanent thermal conditions *steady thermal conditions*	**régimen (térmico) permanente**	régime (thermique) permanent *régime (thermique) stationnaire* ○
1.2.1-53	real fluid	**fluido real**	fluide réel
1.2.1-54	real gas	**gas real**	gaz réel
1.2.1-55	reversible process	**transformación reversible**	transformation réversible
1.2.1-56	second law (of thermodynamics) *second principle (of thermodynamics)* ○	**segundo principio (de la termodinámica)** principio de evolución	second principe (de la thermodynamique)
1.2.1-57	second-law efficiency *exergetic efficiency*	**rendimiento exergético**	rendement exergétique
1.2.1-58	second principle (of thermodynamics) ○ *second law (of thermodynamics)*	**segundo principio (de la termodinámica)** principio de evolución	second principe (de la thermodynamique)
1.2.1-59	stationary state ○ *steady state*	**régimen permanente** régimen estable	régime permanent *régime établi* ○ *régime stable* ○
1.2.1-60	statistical thermodynamics	**termodinámica estadística**	thermodynamique statistique
1.2.1-61	steady state *stationary state* ○	**régimen permanente** régimen estable	régime permanent *régime établi* ○ *régime stable* ○
1.2.1-62	steady thermal conditions *permanent thermal conditions*	**régimen (térmico) permanente**	régime (thermique) permanent *régime (thermique) stationnaire* ○
1.2.1-63	surroundings	**medio exterior al sistema termodinámico**	milieu extérieur

ENGLISH	ESPAÑOL	FRANÇAIS	
(thermal) equation of state	**ecuación de estado**	équation d'état (thermique)	1.2.1-**64**
thermal equilibrium	**equilibrio térmico**	équilibre thermique	1.2.1-**65**
thermodynamic equilibrium	**equilibrio termodinámico**	équilibre thermodynamique	1.2.1-**66**
thermodynamic similarity	**similitud termodinámica** semejanza termodinámica	similitude thermodynamique	1.2.1-**67**
(thermodynamic) system	**sistema termodinámico**	système (thermodynamique)	1.2.1-**68**
thermodynamics	**termodinámica**	thermodynamique	1.2.1-**69**
thermophysics	**termofísica**	thermophysique	1.2.1-**70**
third law (of thermodynamics) *third principle (of thermodynamics)* ◐	**tercer principio (de la termodinámica)**	troisième principe (de la thermodynamique)	1.2.1-**71**
third principle (of thermodynamics) ◐ *third law (of thermodynamics)*	**tercer principio (de la termodinámica)**	troisième principe (de la thermodynamique)	1.2.1-**72**
transient state	**régimen transitorio**	régime transitoire	1.2.1-**73**
transport property	**propiedades de transporte**	propriété de transport	1.2.1-**74**
unsteady state	**régimen variable**	régime variable	1.2.1-**75**
virial coefficients	**coeficientes del virial** coeficientes viriales	coefficients du viriel	1.2.1-**76**
zero point energy	**energía de punto cero**	énergie de point zéro	1.2.1-**77**
zero principle *zeroth law of thermodynamics*	**principio cero**	principe zéro	1.2.1-**78**
zeroth law of thermodynamics *zero principle*	**principio cero**	principe zéro	1.2.1-**79**

SECTION 1.2.2 *Phase change*	SUBCAPÍTULO 1.2.2 *Cambio de fase*	SOUS-CHAPITRE 1.2.2 *Changement de phase*	
atmospheric boiling point *normal boiling point*	**punto de ebullición**	point d'ébullition *température d'ébullition*	1.2.2-**1**
azeotrope *azeotropic mixture*	**azeótropo** mezcla azeotrópica	azéotrope *mélange azéotropique* ◐	1.2.2-**2**
azeotropic	**azeótropico**	azéotrope (adj.) *azéotropique*	1.2.2-**3**
azeotropic mixture *azeotrope*	**azeótropo** mezcla azeotrópica	azéotrope *mélange azéotropique* ◐	1.2.2-**4**
azeotropic point	**punto azeotrópico**	point azéotropique *point d'azéotropie* ◐	1.2.2-**5**
azeotropy	**azeotropía**	azéotropie	1.2.2-**6**
boiling *ebullition* ◐	**ebullición**	ébullition	1.2.2-**7**
bubble point	**punto de burbuja**	point de bulle	1.2.2-**8**
burnout point *critical (nucleate boiling) heat flux* *maximum nucleate boiling heat flux* *peak nucleate boiling heat flux*	**flujo máximo de ebullición nucleada** flujo crítico de ebullición nucleada	flux maximal de l'ébullition nucléée *flux critique de l'ébullition nucléée*	1.2.2-**9**
change of state (1)	**cambio de estado**	changement d'état	1.2.2-**10**
change of state (2) ◐ *phase change*	**cambio de fase** cambio de estado	changement de phase *transition de phase*	1.2.2-**11**

	ENGLISH	ESPAÑOL	FRANÇAIS
1.2.2-**12**	chemical potential	**potencial químico**	potentiel chimique
1.2.2-**13**	condensate	**condensado**	condensat *produit de condensation* ○
1.2.2-**14**	condensation *liquefaction*	**condensación** licuación licuefacción	condensation *liquéfaction* ○
1.2.2-**15**	condensation point	**punto de condensación**	point de condensation
1.2.2-**16**	critical (nucleate boiling) heat flux *burnout point* *maximum nucleate boiling heat flux* *peak nucleate boiling heat flux*	**flujo máximo de ebullición nucleada** flujo crítico de ebullición nucleada	flux maximal de l'ébullition nucléée *flux critique de l'ébullition nucléée*
1.2.2-**17**	critical point *critical state*	**punto crítico** estado crítico	état critique
1.2.2-**18**	critical pressure	**presión crítica**	pression critique
1.2.2-**19**	critical state *critical point*	**punto crítico** estado crítico	état critique
1.2.2-**20**	critical temperature	**temperatura crítica**	température critique
1.2.2-**21**	critical volume	**volumen crítico**	volume critique
1.2.2-**22**	cryoscopy	**crioscopia** criometría	cryoscopie *cryométrie*
1.2.2-**23**	depression of freezing point	**disminución del punto de congelación**	abaissement du point de congélation
1.2.2-**24**	dew point	**punto de rocío**	point de rosée
1.2.2-**25**	droplet condensation	**condensación en gotas**	condensation en gouttes
1.2.2-**26**	(dry) saturated vapour	**vapor saturado (seco)**	vapeur saturante *vapeur saturée (sèche)*
1.2.2-**27**	ebullition ○ *boiling*	**ebullición**	ébullition
1.2.2-**28**	eutectic	**eutéctico**	eutectique
1.2.2-**29**	eutectic point	**punto eutéctico**	point d'eutexie *point eutectique* ○
1.2.2-**30**	eutexy	**eutexia**	eutexie
1.2.2-**31**	evaporation	**evaporación**	évaporation
1.2.2-**32**	evaporation rate	**caudal de evaporación**	débit d'évaporation
1.2.2-**33**	film boiling	**ebullición en película**	ébullition en film *ébullition pelliculaire* ○
1.2.2-**34**	film condensation	**condensación en película**	condensation en film *condensation pelliculaire* ○
1.2.2-**35**	fluid	**fluido**	fluide
1.2.2-**36**	freezing	**congelación**	congélation
1.2.2-**37**	freezing point	**punto de congelación**	point de congélation *point de solidification* ○
1.2.2-**38**	ice melting equivalent	**punto de fusión del hielo**	chaleur latente de la glace
1.2.2-**39**	ice (melting) point	**calor latente del hielo**	point de (fusion de la) glace *température de la glace fondante* ○
1.2.2-**40**	latent heat	**calor latente**	chaleur latente
1.2.2-**41**	liquefaction *condensation*	**condensación** licuación licuefacción	condensation *liquéfaction* ○

| --- | --- | --- | --- |
liquefaction of gases	**licuación de (los) gases** licuefacción de (los) gases	liquéfaction des gaz	1.2.2-**42**
liquid-vapour mixture	**mezcla líquido-vapor**	mélange liquide-vapeur	1.2.2-**43**
maximum nucleate boiling heat flux *burnout point* *critical (nucleate boiling) heat flux* *peak nucleate boiling heat flux*	**flujo máximo de ebullición nucleada** flujo crítico de ebullición nucleada	flux maximal de l'ébullition nucléée *flux critique de l'ébullition nucléée*	1.2.2-**44**
melting	**fusión**	fusion	1.2.2-**45**
melting point	**punto de fusión**	point de fusion	1.2.2-**46**
non-azeotropic mixture *zeotrope* *zeotropic mixture*	**zeótropo** mezcla zeotrópica mezcla no-azeotrópica	zéotrope *mélange non-azéotropique* ◑ *mélange zéotropique* ◑	1.2.2-**47**
normal boiling point *atmospheric boiling point*	**punto de ebullición**	point d'ébullition *température d'ébullition*	1.2.2-**48**
nucleate boiling	**ebullición nucleada**	ébullition nucléée	1.2.2-**49**
nucleation	**nucleación**	nucléation	1.2.2-**50**
peak nucleate boiling heat flux *burnout point* *critical (nucleate boiling) heat flux* *maximum nucleate boiling heat flux*	**flujo máximo de ebullición nucleada** flujo crítico de ebullición nucleada	flux maximal de l'ébullition nucléée *flux critique de l'ébullition nucléée*	1.2.2-**51**
phase change *change of state (2)* ◑	**cambio de fase** cambio de estado	changement de phase *transition de phase*	1.2.2-**52**
pool boiling	**ebullición libre**	ébullition libre	1.2.2-**53**
quality	**título de vapor**	titre en vapeur	1.2.2-**54**
saturated liquid	**líquido saturado**	liquide saturant	1.2.2-**55**
saturated vapour pressure	**presión de vapor saturado** tensión de vapor saturado	pression de vapeur saturante	1.2.2-**56**
saturation	**saturación**	saturation	1.2.2-**57**
saturation pressure	**presión de saturación**	pression de saturation	1.2.2-**58**
saturation temperature	**temperatura de saturación**	température de saturation	1.2.2-**59**
solidification	**solidificación**	solidification	1.2.2-**60**
solidification point	**punto de solidificación**	point de solidification	1.2.2-**61**
subcooled liquid	**líquido subenfriado**	liquide sous-refroidi	1.2.2-**62**
subcooling	**subenfriamiento**	sous-refroidissement	1.2.2-**63**
sublimation (desublimation)	**sublimación**	sublimation (désublimation)	1.2.2-**64**
(suction) superheat	**sobrecalentamiento del vapor aspirado** sobrecalentamiento de succión	surchauffe à l'aspiration	1.2.2-**65**
supercooling	**sobrefusión**	surfusion	1.2.2-**66**
superheat	**sobrecalentamiento**	surchauffe	1.2.2-**67**
superheat (to)	**sobrecalentar**	surchauffer	1.2.2-**68**
superheated vapour	**vapor sobrecalentado**	vapeur surchauffée	1.2.2-**69**
supersaturated vapour	**vapor sobresaturado**	vapeur sursaturée	1.2.2-**70**
supersaturation	**sobresaturación**	sursaturation	1.2.2-**71**
triple point	**punto triple**	point triple	1.2.2-**72**
vaporization	**vaporización**	vaporisation	1.2.2-**73**
vapour	**vapor**	vapeur	1.2.2-**74**

ENGLISH	ESPAÑOL	FRANÇAIS
1.2.2-75 wet vapour	**vapor húmedo**	vapeur humide
1.2.2-76 zeotrope *non-azeotropic mixture* *zeotropic mixture*	**zeótropo** mezcla zeotrópica mezcla no-azeotrópica	zéotrope *mélange non-azéotropique* ● *mélange zéotropique* ●
1.2.2-77 zeotropic mixture *non-azeotropic mixture* *zeotrope*	**zeótropo** mezcla zeotrópica mezcla no-azeotrópica	zéotrope *mélange non-azéotropique* ● *mélange zéotropique* ●

SECTION 1.2.3 *Thermodynamic processes*	SUBCAPÍTULO 1.2.3 *Transformaciones termodinámicas*	SOUS-CHAPITRE 1.2.3 *Transformations thermodynamiques*
1.2.3-1 absorption (refrigeration) cycle	**ciclo (frigorífico) de absorción** ciclo de compresión térmica	cycle (frigorifique) à absorption
1.2.3-2 actual cycle	**ciclo real**	cycle réel
1.2.3-3 adiabatic	**adiabático**	adiabatique
1.2.3-4 adiabatic compression	**compresión adiabática**	compression adiabatique
1.2.3-5 adiabatic expansion	**expansión adiabática**	détente adiabatique
1.2.3-6 adiabatic exponent	**exponente adiabático**	exposant adiabatique
1.2.3-7 air (refrigeration) cycle	**ciclo (frigorífico) de aire**	cycle (frigorifique) à air
1.2.3-8 Carnot cycle	**ciclo de Carnot**	cycle de Carnot
1.2.3-9 closed cycle	**ciclo cerrado**	cycle en circuit "fermé"
1.2.3-10 compressibility	**compresibilidad**	compressibilité
1.2.3-11 compression	**compresión**	compression
1.2.3-12 compression exponent *compression index*	**exponente de compresión**	exposant de compression
1.2.3-13 compression index *compression exponent*	**exponente de compresión**	exposant de compression
1.2.3-14 compression (refrigeration) cycle	**ciclo (frigorífico) de compresión** ciclo de compresión mecánica	cycle (frigorifique) à compression
1.2.3-15 degrees of freedom *degrees of liberty* ○	**grados de libertad**	degrés de liberté *variance*
1.2.3-16 degrees of liberty ○ *degrees of freedom*	**grados de libertad**	degrés de liberté *variance*
1.2.3-17 enthalpy chart *enthalpy diagram*	**diagrama entálpico**	diagramme enthalpique
1.2.3-18 enthalpy diagram *enthalpy chart*	**diagrama entálpico**	diagramme enthalpique
1.2.3-19 entropy chart *entropy diagram*	**diagrama entrópico**	diagramme entropique
1.2.3-20 entropy diagram *entropy chart*	**diagrama entrópico**	diagramme entropique
1.2.3-21 expander cycle *work cycle* *work extraction cycle*	**ciclo con trabajo exterior**	cycle avec travail extérieur
1.2.3-22 expansion	**expansión**	détente
1.2.3-23 indicator diagram	**diagrama indicador**	diagramme indicateur

ENGLISH	ESPAÑOL	FRANÇAIS	
isenthalp	**isoentálpica** (subst.) linea isoentálpica	isenthalpe	1.2.3-**24**
isenthalpic	**isoentálpico** (adj.)	isenthalpique	1.2.3-**25**
isenthalpic expansion	**expansión isoentálpica**	détente isenthalpique	1.2.3-**26**
isentrope	**isoentrópica** (subst.)	isentrope	1.2.3-**27**
isentropic	**isoentrópico** (adj.)	isentropique	1.2.3-**28**
isobar	**isóbara** (subst.)	isobare (subst.)	1.2.3-**29**
isobaric	**isobárico** (adj.) isóbaro (adj.)	isobare (adj.)	1.2.3-**30**
isochor	**isócora** (subst.)	isochore (subst.)	1.2.3-**31**
isochoric	**isócoro** (adj.)	isochore (adj.) *isovolume* ○	1.2.3-**32**
isotherm	**isoterma** (subst.)	isotherme (subst.)	1.2.3-**33**
isothermal	**isotérmico** (adj.) isotermo (adj.)	isotherme (adj.) *isothermique* ○	1.2.3-**34**
(isothermal) compressibility	**compresibilidad**	compressibilité (isotherme)	1.2.3-**35**
Joule-Thomson effect	**efecto Joule-Thomson**	effet Joule-Thomson	1.2.3-**36**
Joule-Thomson process	**transformación Joule-Thomson**	détente sans travail extérieur	1.2.3-**37**
Lorenz cycle	**ciclo de Lorentz**	cycle de Lorenz	1.2.3-**38**
Mollier chart *Mollier diagram*	**diagrama de Mollier**	diagramme de Mollier	1.2.3-**39**
Mollier diagram *Mollier chart*	**diagrama de Mollier**	diagramme de Mollier	1.2.3-**40**
open cycle	**ciclo abierto**	cycle ouvert	1.2.3-**41**
polytrop	**politrópica** (subst.)	polytrope	1.2.3-**42**
polytropic	**politrópico** (adj.)	polytropique	1.2.3-**43**
polytropic compression	**compresión politrópica**	compression polytropique	1.2.3-**44**
polytropic expansion	**expansión politrópica**	détente polytropique	1.2.3-**45**
polytropic exponent	**exponente politrópico**	exposant polytropique	1.2.3-**46**
polytropic process	**proceso politrópico**	processus polytropique	1.2.3-**47**
pressure volume chart *PV chart* *pressure volume diagram*	**diagrama presión-volumen** diagrama dinámico diagrama de Clapeyron	diagramme de Clapeyron *diagramme pression volume* *diagramme PV* ○	1.2.3-**48**
pressure volume diagram *PV chart* *pressure volume chart*	**diagrama presión-volumen** diagrama dinámico diagrama de Clapeyron	diagramme de Clapeyron *diagramme pression volume* *diagramme PV* ○	1.2.3-**49**
PV chart *pressure volume chart* *pressure volume diagram*	**diagrama presión-volumen** diagrama dinámico diagrama de Clapeyron	diagramme de Clapeyron *diagramme pression volume* *diagramme PV* ○	1.2.3-**50**
Rankine cycle	**ciclo de Rankine**	cycle de Rankine	1.2.3-**51**
refrigeration cycle	**ciclo frigorífico**	cycle frigorifique	1.2.3-**52**
reverse cycle	**ciclo inverso**	cycle inversé	1.2.3-**53**
reversible cycle	**ciclo reversible**	cycle réversible	1.2.3-**54**
semi-closed cycle	**ciclo semicerrado**	cycle semi-fermé	1.2.3-**55**
standard rating cycle	**ciclo de referencia**	cycle de référence	1.2.3-**56**

	ENGLISH	ESPAÑOL	FRANÇAIS
1.2.3-**57**	state diagram	**diagrama de estado**	diagramme d'état
1.2.3-**58**	steam jet (refrigeration) cycle *vapour jet (refrigeration) cycle*	**ciclo (frigorífico) de eyección (de vapor)**	cycle (frigorifique) à éjection (de vapeur)
1.2.3-**59**	Stirling cycle	**ciclo de Stirling**	cycle de Stirling
1.2.3-**60**	(thermodynamic) cycle	**ciclo termodinámico**	cycle (thermodynamique)
1.2.3-**61**	throttling expansion	**expansión sin trabajo externo**	détente sans travail extérieur
1.2.3-**62**	vapour jet (refrigeration) cycle *steam jet (refrigeration) cycle*	**ciclo (frigorífico) de eyección (de vapor)**	cycle (frigorifique) à éjection (de vapeur)
1.2.3-**63**	work cycle *expander cycle* *work extraction cycle*	**ciclo con trabajo exterior**	cycle avec travail extérieur
1.2.3-**64**	work extraction cycle *expander cycle* *work cycle*	**ciclo con trabajo exterior**	cycle avec travail extérieur
1.2.3-**65**	working fluid *working substance*	**fluido activo** fluido de trabajo	fluide actif
1.2.3-**66**	working substance *working fluid*	**fluido activo** fluido de trabajo	fluide actif

	SECTION 1.3 *Heat and mass transfer* SECTION 1.3.1 *Heat and mass transfer:* *general background*	SUBCAPÍTULO 1.3 *Transmisión de calor y de masa* SUBCAPÍTULO 1.3.1 *Generalidades sobre la transmisión* *de calor y de masa*	SOUS-CHAPITRE 1.3 *Transfert de chaleur et de masse* SOUS-CHAPITRE 1.3.1 *Généralités sur le transfert de chaleur* *et de masse*
1.3.1-**1**	cooling range	**amplitud del enfriamiento**	amplitude du refroidissement
1.3.1-**2**	energy flow rate	**flujo energético**	flux énergétique
1.3.1-**3**	film coefficient of heat transfer ○ *heat transfer coefficient* *surface coefficient of heat transfer* ○ *surface film conductance* ○	**coeficiente de transmisión térmica** coeficiente de convección térmica coeficiente de transmisión superficial coeficiente de película	coefficient de transfert de chaleur *coefficient d'échange (thermique) superficiel* *coefficient de transmission (thermique)* *de surface*
1.3.1-**4**	heat	**calor**	chaleur
1.3.1-**5**	heat exchange	**intercambio de calor** intercambio térmico	échange thermique *échange de chaleur* ○
1.3.1-**6**	heat flow	**flujo de calor** paso de calor	transport thermique *transport de chaleur* ○
1.3.1-**7**	heat flow meter	**flujómetro térmico** medidor de flujo de calor	fluxmètre thermique
1.3.1-**8**	heat flow path	**línea de flujo térmico**	ligne de flux thermique *ligne de flux de chaleur* ○
1.3.1-**9**	heat flow rate	**flujo de calor** flujo térmico	flux thermique
1.3.1-**10**	heat lag ○ *thermal lag*	**desfase en la transmisión del calor**	temps mort thermique
1.3.1-**11**	heat transfer *heat transmission* ○ *heat transport* ○ *thermal transmission* ○	**transmisión de calor** transmisión térmica	transfert de chaleur *transmission de chaleur* ○

ENGLISH	ESPAÑOL	FRANÇAIS	
heat transfer coefficient *film coefficient of heat transfer* ☉ *surface coefficient of heat transfer* ☉ *surface film conductance* ☉	**coeficiente de transmisión térmica** coeficiente de convección térmica coeficiente de transmisión superficial coeficiente de película	coefficient de transfert de chaleur *coefficient d'échange (thermique) superficiel* *coefficient de transmission (thermique)* *de surface*	1.3.1-**12**
heat transmission ☉ *heat transfer* *heat transport* ☉ *thermal transmission* ☉	**transmisión de calor** transmisión térmica	transfert de chaleur *transmission de chaleur* ☉	1.3.1-**13**
heat transport ☉ *heat transfer* *heat transmission* ☉ *thermal transmission* ☉	**transmisión de calor** transmisión térmica	transfert de chaleur *transmission de chaleur* ☉	1.3.1-**14**
overall heat transfer coefficient *thermal transmittance*	**transmitancia térmica** coeficiente global de transmisión de calor coeficiente global de transmisión térmica	coefficient global de transfert de chaleur *coefficient global d'échange thermique* *transmittance thermique* ☉	1.3.1-**15**
surface coefficient of heat transfer ☉ *heat transfer coefficient* *film coefficient of heat transfer* ☉ *surface film conductance* ☉	**coeficiente de transmisión térmica** coeficiente de convección térmica coeficiente de transmisión superficial coeficiente de película	coefficient de transfert de chaleur *coefficient d'échange (thermique) superficiel* *coefficient de transmission (thermique)* *de surface*	1.3.1-**16**
surface film conductance ☉ *heat transfer coefficient* *film coefficient of heat transfer* ☉ *surface coefficient of heat transfer* ☉	**coeficiente de transmisión térmica** coeficiente de convección térmica coeficiente de transmisión superficial coeficiente de película	coefficient de transfert de chaleur *coefficient d'échange (thermique) superficiel* *coefficient de transmission (thermique)* *de surface*	1.3.1-**17**
temperature field	**campo térmico** campo de temperaturas	champ de température *champ thermique*	1.3.1-**18**
thermal boundary resistance *thermal contact resistance*	**resistencia térmica límite** resistencia térmica de contacto	résistance thermique de contact *résistance thermique limite* ☉	1.3.1-**19**
thermal contact resistance *thermal boundary resistance*	**resistencia térmica límite** resistencia térmica de contacto	résistance thermique de contact *résistance thermique limite* ☉	1.3.1-**20**
thermal lag *heat lag* ☉	**desfase en la transmisión del calor**	temps mort thermique	1.3.1-**21**
thermal resistance	**resistencia térmica** coeficiente de resistencia térmica	résistance thermique	1.3.1-**22**
thermal transmission ☉ *heat transfer* *heat transmission* ☉ *heat transport* ☉	**transmisión de calor** transmisión térmica	transfert de chaleur *transmission de chaleur* ☉	1.3.1-**23**
thermal transmittance *overall heat transfer coefficient*	**transmitancia térmica** coeficiente global de transmisión de calor coeficiente global de transmisión térmica	coefficient global de transfert de chaleur *coefficient global d'échange thermique* *transmittance thermique* ☉	1.3.1-**24**
transient heat flow	**flujo de calor en régimen transitorio**	écoulement de chaleur en régime transitoire	1.3.1-**25**

SECTION 1.3.2 *Radiation*	**SUBCAPÍTULO 1.3.2** *Radiación*	**SOUS-CHAPITRE 1.3.2** *Rayonnement*	
absorptance *absorptivity* *absorption factor* ☉	**absorbancia** coeficiente de absorción poder absorbente	absorptance *absorptivité* *facteur d'absorption*	1.3.2-**1**
absorption coefficient	**coeficiente de absorcion**	coefficient d'absorption	1.3.2-**2**
absorption factor ☉ *absorptance* *absorptivity*	**absorbancia** coeficiente de absorción poder absorbente	absorptance *absorptivité* *facteur d'absorption*	1.3.2-**3**

ENGLISH	ESPAÑOL	FRANÇAIS
1.3.2-4 absorptivity *absorptance* *absorption factor* ◐	**absorbancia** coeficiente de absorción poder absorbente	absorptance *absorptivité* *facteur d'absorption*
1.3.2-5 angle factor *shape factor* *view factor* *configuration factor* ◐	**factor de angulo** factor de forma	facteur d'angle *facteur de forme*
1.3.2-6 athermanous	**atérmano**	athermane
1.3.2-7 black body	**cuerpo negro**	corps noir
1.3.2-8 configuration factor ◐ *angle factor* *shape factor* *view factor*	**factor de angulo** factor de forma	facteur d'angle *facteur de forme*
1.3.2-9 diathermanous	**diatérmano**	diathermane
1.3.2-10 diffuse surface	**superficie difusa**	surface diffusante
1.3.2-11 emissive power *emittance* *radiant exitance*	**emitancia** exitancia radiante	émittance *exitance*
1.3.2-12 emissivity	**emisividad (total)**	émissivité
1.3.2-13 emittance *emissive power* *radiant exitance*	**emitancia** exitancia radiante	émittance *exitance*
1.3.2-14 gray body (USA) *grey body*	**cuerpo gris**	corps gris
1.3.2-15 grey body *gray body (USA)*	**cuerpo gris**	corps gris
1.3.2-16 interstellar cooling *sky cooling*	**enfriamiento por radiación terrestre**	rayonnement sur l'espace *refroidissement par rayonnement terrestre*
1.3.2-17 mean radiant temperature	**temperatura media radiante**	température radiante moyenne
1.3.2-18 monochromatic emissivity ◐ *spectral emissivity*	**emisividad espectral** emisividad monocromática	émissivité spectrale *émissivité monochromatique* ◐
1.3.2-19 radiance *radiant intensity*	**luminancia** radiancia	luminance énergétique
1.3.2-20 radiant exitance *emissive power* *emittance*	**emitancia** exitancia radiante	émittance *exitance*
1.3.2-21 radiant heat	**calor radiante**	chaleur radiante *chaleur rayonnante*
1.3.2-22 radiant intensity *radiance*	**luminancia** radiancia	luminance énergétique
1.3.2-23 radiation	**radiación**	rayonnement
1.3.2-24 radiation heat transfer coefficient	**coeficiente de radiación térmica**	coefficient d'échange thermique radiatif
1.3.2-25 radiation shield	**pantalla antirradiación**	écran antirayonnement
1.3.2-26 radiometer	**radiómetro**	radiomètre
1.3.2-27 reflectance *reflectivity* *reflection factor* ◐	**reflectancia** coeficiente de reflexión poder reflector	facteur de réflexion *réflectance* *réflectivité*
1.3.2-28 reflection factor ◐ *reflectance* *reflectivity*	**reflectancia** coeficiente de reflexión poder reflector	facteur de réflexion *réflectance* *réflectivité*

ENGLISH	ESPAÑOL	FRANÇAIS	
reflectivity *reflectance* *reflection factor* ◑	**reflectancia** coeficiente de reflexión poder reflector	facteur de réflexion *réflectance* *réflectivité*	1.3.2-**29**
shape factor *angle factor* *view factor* *configuration factor* ◑	**factor de angulo** factor de forma	facteur d'angle *facteur de forme*	1.3.2-**30**
sky cooling *interstellar cooling*	**enfriamiento por radiación terrestre**	rayonnement sur l'espace *refroidissement par rayonnement terrestre*	1.3.2-**31**
solar constant	**constante solar**	constante solaire	1.3.2-**32**
spectral	**espectral**	spectral	1.3.2-**33**
spectral emissivity *monochromatic emissivity* ◑	**emisividad espectral** emisividad monocromática	émissivité spectrale *émissivité monochromatique* ◑	1.3.2-**34**
spectral emittance *spectral exitance*	**exitancia espectral** emitancia monocromática	exitance spectrale *émittance spectrale* ◑	1.3.2-**35**
spectral exitance *spectral emittance*	**exitancia espectral** emitancia monocromática	exitance spectrale *émittance spectrale* ◑	1.3.2-**36**
transmission factor *transmissivity* *transmittance*	**transmitancia** coeficiente de transmisión poder transmitido	transmittance *transmittivité* *facteur de transmission*	1.3.2-**37**
transmissivity *transmission factor* *transmittance*	**transmltancia** coeficiente de transmisión poder transmitido	transmittance *transmittivité* *facteur de transmission*	1.3.2-**38**
transmittance *transmission factor* *transmissivity*	**transmitancia** coeficiente de transmisión poder transmitido	transmittance *transmittivité* *facteur de transmission*	1.3.2-**39**
view factor *angle factor* *shape factor* *configuration factor* ◑	**factor de angulo** factor de forma	facteur d'angle *facteur de forme*	1.3.2-**40**

SECTION 1.3.3 *Thermal conduction*	SUBCAPÍTULO 1.3.3 *Conducción térmica*	SOUS-CHAPITRE 1.3.3 *Conduction thermique*	
heat conduction *thermal conduction* ◑	**conducción térmica**	conduction thermique	1.3.3-**1**
thermal conduction ◑ *heat conduction*	**conducción de calor**	conduction thermique	1.3.3-**2**
thermal conductivity	**conductividad térmica** coeficiente de conductibilidad térmica	conductivité thermique *coefficient de conductibilité thermique* ○ *coefficient de conduction thermique* ○	1.3.3-**3**
thermal diffusivity	**difusividad térmica** difusión térmica	diffusivité thermique	1.3.3-**4**
thermal resistivity	**resistividad térmica**	résistivité thermique	1.3.3-**5**

SECTION 1.3.4 *Convection*	SUBCAPÍTULO 1.3.4 *Convección*	SOUS-CHAPITRE 1.3.4 *Convection*	
convection (of heat)	**convección de calor** convección térmica	convection (de chaleur) *transfert convectif*	1.3.4-**1**

	ENGLISH	ESPAÑOL	FRANÇAIS
1.3.4-**2**	convector (fluid)	**fluido convector**	fluide convecteur
1.3.4-**3**	film coefficient of heat transfer ○ *heat transfer coefficient* *surface coefficient of heat transfer* ○ *surface film conductance* ○	**coeficiente de transmissión térmica** coeficiente de convección térmica coeficiente de transmisión superficial coeficiente de película	coefficient de transfert de chaleur *coefficient d'échange (thermique) superficiel* *coefficient de transmission (thermique)* *de surface*
1.3.4-**4**	forced convection	**convección forzada**	convection forcée
1.3.4-**5**	free convection *natural convection*	**convección natural** convección libre	convection naturelle *convection libre*
1.3.4-**6**	heat transfer coefficient *film coefficient of heat transfer* ○ *surface coefficient of heat transfer* ○ *surface film conductance* ○	**coeficiente de transmissión térmica** coeficiente de convección térmica coeficiente de transmisión superficial coeficiente de película	coefficient de transfert de chaleur *coefficient d'échange (thermique) superficiel* *coefficient de transmission (thermique)* *de surface*
1.3.4-**7**	natural convection *free convection*	**convección natural** convección libre	convection naturelle *convection libre*
1.3.4-**8**	surface coefficient of heat transfer ○ *heat transfer coefficient* *film coefficient of heat transfer* ○ *surface film conductance* ○	**coeficiente de transmissión térmica** coeficiente de convección térmica coeficiente de transmisión superficial coeficiente de película	coefficient de transfert de chaleur *coefficient d'échange (thermique) superficiel* *coefficient de transmission thermique* *de surface*
1.3.4-**9**	surface film conductance ○ *heat transfer coefficient* *film coefficient of heat transfer* ○ *surface coefficient of heat transfer* ○	**coeficiente de transmissión térmica** coeficiente de convección térmica coeficiente de transmisión superficial coeficiente de película	coefficient de transfert de chaleur *coefficient d'échange (thermique) superficiel* *coefficient de transmission thermique* *de surface*

	SECTION 1.3.5 *Mass transfer*	SUBCAPÍTULO 1.3.5 *Transporte de masa*	SOUS-CHAPITRE 1.3.5 *Transfert de masse*
1.3.5-**1**	accommodation coefficient	**coeficiente de acomodación**	coefficient d'accommodation
1.3.5-**2**	concentration	**concentración**	concentration
1.3.5-**3**	convective diffusivity *convective mass transfer* *eddy diffusion* *turbulent diffusion*	**difusión convectiva**	diffusion convective
1.3.5-**4**	convective mass transfer *convective diffusivity* *eddy diffusion* *turbulent diffusion*	**difusión convectiva**	diffusion convective
1.3.5-**5**	diffusion *molecular diffusion*	**difusión** difusión molecular	diffusion
1.3.5-**6**	diffusion coefficient *mass diffusivity*	**coeficiente de difusión**	coefficient de diffusion (de masse)
1.3.5-**7**	eddy diffusion *convective diffusivity* *convective mass transfer* *turbulent diffusion*	**difusión convectiva**	diffusion convective
1.3.5-**8**	mass diffusivity *diffusion coefficient*	**coeficiente de difusión**	coefficient de diffusion (de masse)
1.3.5-**9**	mass transfer *mass transport* ○	**transporte de masa** transmisión de masa	transfert de masse *transport de masse* ○
1.3.5-**10**	mass transport ○ *mass transfer*	**transporte de masa** transmisión de masa	transfert de masse *transport de masse* ○
1.3.5-**11**	molecular diffusion *diffusion*	**difusión** difusión molecular	diffusion

ENGLISH	ESPAÑOL	FRANÇAIS	
solute	**soluto**	soluté	1.3.5-**12**
solvent	**disolvente**	solvant	1.3.5-**13**
turbulent diffusion *convective diffusivity* *convective mass transfer* *eddy diffusion*	**difusión convectiva**	diffusion convective	1.3.5-**14**

SECTION 1.4 *Related fields* SECTION 1.4.1 *Acoustics*	SUBCAPÍTULO 1.4 *Campos asociados* SUBCAPÍTULO 1.4.1 *Acústica*	SOUS-CHAPITRE 1.4 *Domaines apparentés* SOUS-CHAPITRE 1.4.1 *Acoustique*	
acoustic oscillation *sound* *acoustic vibration* ◐	**oscilación acústica** sonido vibración acústica	oscillation acoustique *son* *vibration acoustique* ◐	1.4.1-**1**
acoustic vibration ◐ *acoustic oscillation* *sound*	**oscilación acústica** sonido vibración acústica	oscillation acoustique *son* *vibration acoustique* ◐	1.4.1-**2**
ambient noise	**ruido ambiental** ruido ambiente	bruit ambiant	1.4.1-**3**
audible sound	**sonido audible**	son audible	1.4.1-**4**
background noise	**ruido de fondo**	bruit de fond	1.4.1-**5**
hertz (Hz)	**hercio (Hz)**	hertz (Hz)	1.4.1-**6**
infrasound	**infrasonido**	infrason	1.4.1-**7**
instantaneous sound pressure	**presión de sonido instantánea**	pression acoustique instantanée	1.4.1-**8**
noise	**ruido**	bruit	1.4.1-**9**
octave band	**banda de octava**	bande d'octave	1.4.1-**10**
one-third-octave band	**banda de un tercio de octava**	bande de tiers d'octave	1.4.1-**11**
pink noise	**sonido rosa**	bruit rose	1.4.1-**12**
rms sound pressure	**raíz cuadrática media de la presión de sonido** presión de sonido cuadrática media	pression acoustique efficace	1.4.1-**13**
sound *acoustic oscillation* *acoustic vibration* ◐	**oscilación acústica** sonido vibración acústica	oscillation acoustique *son* *vibration acoustique* ◐	1.4.1-**14**
sound power level (L_W)	**nivel de potencia sonora (L_W)**	niveau de puissance acoustique (L_W)	1.4.1-**15**
sound power of a source	**potencia sonora de una fuente**	puissance acoustique d'une source	1.4.1-**16**
sound pressure	**presión de sonido**	pression acoustique	1.4.1-**17**
sound pressure level	**nivel de presión sonora** nivel de presión de sonido	niveau de pression acoustique	1.4.1-**18**
sound spectrum	**espectro de sonido**	spectre acoustique	1.4.1-**19**
static pressure	**presión estática**	pression statique	1.4.1-**20**
ultrasound	**ultrasonido**	ultrason	1.4.1-**21**
white noise	**ruido blanco**	bruit blanc	1.4.1-**22**

ENGLISH	ESPAÑOL	FRANÇAIS
SECTION 1.4.2 *Electricity*	**SUBCAPÍTULO 1.4.2** *Electricidad*	**SOUS-CHAPITRE 1.4.2** *Electricité*
1.4.2-**1** absorbed electrical capacity	**potencia eléctrica absorbida**	puissance électrique absorbée
1.4.2-**2** effective power input	**potencia absorbida efectiva**	puissance absorbée effective
1.4.2-**3** rated frequency	**frecuencia nominal** frecuencia especificada	fréquence nominale
1.4.2-**4** rated voltage	**tensión nominal** voltaje especificado	tension nominale
1.4.2-**5** starting current	**corriente de arranque**	intensité de démarrage
1.4.2-**6** total power input	**potencia absorbida total**	puissance absorbée totale

Capítulo 2.

PRODUCCIÓN DE FRÍO

● término aceptado

○ término obsoleto

ENGLISH	ESPAÑOL	FRANÇAIS	
SECTION 2.1 *Heat balance*	**SUBCAPÍTULO 2.1** *Balance térmico*	**SOUS-CHAPITRE 2.1** *Bilan thermique*	
cooling load ◐ *heat load* *refrigeration duty* ◐ *refrigeration load* ◐ *refrigeration requirement* ◐	**carga térmica** necesidades de frío	charge thermique *besoin de froid* ◐	2.1-**1**
dry tons (USA) ◐ *sensible heat load*	**carga térmica sensible**	charge thermique "sensible" *charge thermique due à la chaleur sensible* ◐	2.1-**2**
(estimated) design load	**carga de diseño (estimada)**	charge frigorifique prévisionnelle	2.1-**3**
heat balance	**balance térmico**	bilan frigorifique *bilan thermique* ◐	2.1-**4**
heat load *cooling load* ◐ *refrigeration duty* ◐ *refrigeration load* ◐ *refrigeration requirement* ◐	**carga térmica** necesidades de frío	charge thermique *besoin de froid* ◐	2.1-**5**
latent heat load *moisture tons (USA)* ○ *wet tons* ○	**carga térmica latente**	charge thermique "latente" *charge thermique due à la chaleur latente* ◐	2.1-**6**
load factor	**factor de carga**	pourcentage de charge *facteur de charge*	2.1-**7**
moisture tons (USA) ○ *latent heat load* *wet tons* ○	**carga térmica latente**	charge thermique "latente" *charge thermique due à la chaleur latente* ◐	2.1-**8**
product load	**carga térmica debida al producto** necesidades de frío debidas al producto	charge thermique due au produit *besoin de froid pour le produit* ◐	2.1-**9**
refrigeration duty ◐ *heat load* *cooling load* ◐ *refrigeration load* ◐ *refrigeration requirement* ◐	**carga térmica** necesidades de frío	charge thermique *besoin de froid* ◐	2.1-**10**
refrigeration load ◐ *heat load* *cooling load* ◐ *refrigeration duty* ◐ *refrigeration requirement* ◐	**carga térmica** necesidades de frío	charge thermique *besoin de froid* ◐	2.1-**11**
refrigeration requirement ◐ *heat load* *cooling load* ◐ *refrigeration duty* ◐ *refrigeration load* ◐	**carga térmica** necesidades de frío	charge thermique *besoin de froid* ◐	2.1-**12**
sensible heat load *dry tons (USA)* ◐	**carga térmica sensible**	charge thermique "sensible" *charge thermique due à la chaleur sensible* ◐	2.1-**13**
service load	**carga térmica de explotación**	charges thermiques d'exploitation *charges thermiques d'utilisation* ◐	2.1-**14**
wet tons ○ *latent heat load* *moisture tons (USA)* ○	**carga térmica latente**	charge thermique "latente" *charge thermique due à la chaleur latente* ◐	2.1-**15**

ENGLISH	ESPAÑOL	FRANÇAIS
SECTION 2.2 *Refrigeration capacity and calculation data*	**SUBCAPÍTULO 2.2** *Potencia frigorífica y elementos de cálculo*	**SOUS-CHAPITRE 2.2** *Puissance frigorifique et éléments de calcul*

	ENGLISH	ESPAÑOL	FRANÇAIS
2.2-1	adiabatic efficiency ○ *indicated efficiency*	**rendimiento indicado**	rendement indiqué
2.2-2	brake horsepower ○ *power input rating* *shaft horsepower* ○ *shaft power* ○	**potencia efectiva** potencia en el eje	puissance (mécanique) effective *puissance sur l'arbre* ○
2.2-3	coefficient of performance (COP) *performance energy ratio* ○	**coeficiente frigorífico** efecto frigorífico coeficiente de prestaciones	coefficient de performance (COP)
2.2-4	condenser duty ○ *condenser heat*	**calor cedido en el condensador** potencia del condensador	chaleur rejetée au condenseur
2.2-5	condenser heat *condenser duty* ○	**calor cedido en el condensador** potencia del condensador	chaleur rejetée au condenseur
2.2-6	condensing unit capacity ○ *overall refrigerating effect*	**potencia frigorífica global** potencia frigorífica total	puissance frigorifique globale
2.2-7	degree of superheat	**grado de sobrecalentamiento** sobrecalentamiento	(degré de) surchauffe
2.2-8	effective efficiency ○ *overall efficiency*	**rendimiento efectivo** rendimiento global	rendement global *rendement effectif*
2.2-9	effective work	**trabajo efectivo**	travail effectif
2.2-10	efficiency	**rendimiento**	rendement *efficacité* ○
2.2-11	external conditions	**régimen exterior (de temperaturas)**	régime (thermique) extérieur
2.2-12	heat extraction rate	**potencia frigorífica útil** velocidad de extracción de calor	transfert de chaleur utile
2.2-13	heat of subcooling	**calor de subenfriamiento**	chaleur de sous-refroidissement
2.2-14	heat recovery capacity	**potencia del recuperador de calor** capacidad de recuperación de calor	puissance thermique de récupération *puissance du récupérateur de chaleur*
2.2-15	heat removed	**calor cedido**	chaleur enlevée
2.2-16	hold-over ○ *thermal storage*	**acumulación de frío**	stockage de froid *accumulation de froid* ○
2.2-17	ice-making capacity	**capacidad de fabricación de hielo** capacidad de producción de hielo	capacité de production de glace
2.2-18	indicated efficiency *adiabatic efficiency* ○	**rendimiento indicado**	rendement indiqué
2.2-19	indicated power	**potencia indicada**	puissance indiquée
2.2-20	indicated work	**trabajo indicado**	travail indiqué
2.2-21	internal conditions	**régimen interior (de temperaturas)**	régime thermique interne
2.2-22	internal efficiency	**rendimiento interno**	rendement interne
2.2-23	internal power	**potencia interna**	puissance interne
2.2-24	isentropic efficiency	**rendimiento isoentrópico**	rendement isentropique
2.2-25	isentropic power	**potencia isoentrópica**	puissance théorique isentropique
2.2-26	isothermal efficiency	**rendimiento isotérmico**	rendement isothermique *rendement isotherme* ○

ENGLISH	ESPAÑOL	FRANÇAIS	
isothermal power	**potencia isotérmica**	puissance théorique isotherme *puissance isothermique* ○	2.2-**27**
mechanical efficiency	**rendimiento mecánico**	rendement mécanique	2.2-**28**
net cooling capacity	**potencia frigorífica neta**	puissance frigorifique nette	2.2-**29**
net refrigerating effect	**potencia frigorífica neta**	effet frigorifique net	2.2-**30**
overall efficiency *effective efficiency* ○	**rendimiento efectivo** rendimiento global	rendement global *rendement effectif*	2.2-**31**
overall refrigerating effect *condensing unit capacity* ○	**potencia frigorífica global** potencia frigorífica total	puissance frigorifique globale	2.2-**32**
packaged compressor power input	**potencia absorbida de un grupo** **motocompresor**	puissance absorbée d'un groupe motocompresseur	2.2-**33**
performance energy ratio ○ *coefficient of performance (COP)*	**coeficiente frigorífico** efecto frigorífico coeficiente de prestaciones	coefficient de performance (COP)	2.2-**34**
power input	**potencia absorbida**	puissance consommée *puissance sur l'arbre* ○	2.2-**35**
power input rating *brake horsepower* ○ *shaft horsepower* ○ *shaft power* ○	**potencia efectiva** potencia en el eje	puissance (mécanique) effective *puissance sur l'arbre* ○	2.2-**36**
rated conditions	**condiciones nominales**	conditions nominales	2.2-**37**
rating under working conditions	**potencia frigorífica efectiva**	puissance frigorifique effective	2.2-**38**
reference conditions	**condiciones de referencia**	conditions de référence	2.2-**39**
refrigerating capacity	**potencia frigorífica** potencia de enfriamiento	puissance frigorifique	2.2-**40**
refrigerating effect per brake horse- power ○	**potencia frigorífica específica**	puissance frigorifique spécifique ○	2.2-**41**
refrigerating effect per unit of swept volume	**producción frigorífica volumétrica**	production frigorifique volumétrique *production volumétrique effective*	2.2-**42**
(refrigeration) output	**producción frigorífica**	production (frigorifique)	2.2-**43**
shaft horsepower ○ *power input rating* *brake horsepower* ○ *shaft power* ○	**potencia efectiva** potencia en el eje	puissance (mécanique) effective *puissance sur l'arbre* ○	2.2-**44**
shaft power ○ *power input rating* *brake horsepower* ○ *shaft horsepower* ○	**potencia efectiva** potencia en el eje	puissance (mécanique) effective *puissance sur l'arbre* ○	2.2-**45**
specific vaporization enthalpy	**entalpía específica de vaporización**	enthalpie massique d'évaporation	2.2-**46**
standard conditions	**condiciones de referencia** condiciones estandarizadas	conditions nominales *conditions de référence*	2.2-**47**
standard rating *total refrigeration* ○	**potencia frigorífica nominal**	puissance (frigorifique) nominale	2.2-**48**
superheat	**calor de sobrecalentamiento** sobrecalentamiento	chaleur de surchauffe	2.2-**49**
temperature of saturated vapour at the discharge of the compressor	**temperatura del vapor saturado a la** **presión de descarga del compresor**	température de vapeur saturée au refoulement du compresseur *température de saturation vapeur à la* *pression de refoulement*	2.2-**50**
theoretical efficiency	**rendimiento teórico**	rendement théorique	2.2-**51**
thermal storage *hold-over* ○	**acumulación de frío**	stockage de froid *accumulation de froid* ○	2.2-**52**

	ENGLISH	ESPAÑOL	FRANÇAIS
2.2-53	total refrigeration o *standard rating*	**potencia frigorífica nominal**	puissance (frigorifique) nominale
2.2-54	useful refrigerating effect	**potencia frigorífica útil**	puissance frigorifique utile
2.2-55	volumetric efficiency	**rendimiento volumétrico**	rendement volumétrique

	SECTION 2.3 *Refrigerating systems* **SECTION 2.3.1** *Refrigerating systems:* *general background*	**SUBCAPÍTULO 2.3** *Sistemas frigoríficos* **SUBCAPÍTULO 2.3.1** *Generalidades sobre* *sistemas frigoríficos*	**SOUS-CHAPITRE 2.3** *Systèmes frigorifiques* **SOUS-CHAPITRE 2.3.1** *Généralités sur les systèmes* *frigorifiques*
2.3.1-1	control cycle *cycle period*	**ciclo de funcionamiento** ciclo de trabajo	cycle de fonctionnement
2.3.1-2	critical charge	**carga operacional mínima** carga de trabajo mínima	charge minimale (opérationnelle)
2.3.1-3	cycle period *control cycle*	**ciclo de funcionamiento** ciclo de trabajo	cycle de fonctionnement
2.3.1-4	direct refrigerating system	**sistema de enfriamiento directo** enfriamiento por expansión directa sistema frigorifero directo	système de refroidissement direct *refroidissement par détente directe*
2.3.1-5	indirect refrigerating system *secondary cooling system*	**sistema de enfriamiento indirecto** sistema de enfriamiento secundario sistema de enfriamiento por transmisión indirecta sistema frigorifero indirecto	système de refroidissement indirect *système secondaire de refroidissement* *système à fluide secondaire* o
2.3.1-6	limited charge system	**sistema de enfriamiento con carga** **limitada**	système à charge contrôlée
2.3.1-7	mobile system	**instalación frigorífica móvil**	système mobile *système embarqué* o
2.3.1-8	refrigerating machine o *refrigerating system* *refrigeration system* *refrigeration machine* o	**sistema frigorífico** máquina frigorífica	système frigorifique *machine frigorifique*
2.3.1-9	refrigerating system *refrigeration system* *refrigerating machine* o *refrigeration machine* o	**sistema frigorífico** máquina frigorífica	système frigorifique *machine frigorifique*
2.3.1-10	refrigeration machine o *refrigerating system* *refrigeration system* *refrigerating machine* o	**sistema frigorífico** máquina frigorífica	système frigorifique *machine frigorifique*
2.3.1-11	refrigeration system *refrigerating system* *refrigerating machine* o *refrigeration machine* o	**sistema frigorífico** máquina frigorífica	système frigorifique *machine frigorifique*
2.3.1-12	secondary cooling system *indirect refrigerating system*	**sistema de enfriamiento indirecto** sistema de enfriamiento secundario sistema de enfriamiento por transmisión indirecta sistema frigorifero indirecto	système de refroidissement indirect *système secondaire de refroidissement* *système à fluide secondaire*
2.3.1-13	unit system o	**grupo frigorífico autónomo**	groupe frigorifique fabriqué et essayé en usine *système frigorifique monobloc préfabriqué*

ENGLISH	ESPAÑOL	FRANÇAIS	
SECTION 2.3.2 *Compression refrigerating systems*	**SUBCAPÍTULO 2.3.2** *Sistemas frigoríficos de compresión*	**SOUS-CHAPITRE 2.3.2** *Systèmes frigorifiques à compression*	
autorefrigerated cascade ○ *integrated cascade* ◐ *mixed refrigerant cascade* ◐	**cascada integrada**	cascade intégrée *cascade incorporée* ◐	2.3.2-**1**
compound compression *two-stage compression*	**compresión compuesta** compresión en dos etapas compresión en doble salto	compression biétagée *compression à deux étages* ◐	2.3.2-**2**
compression ratio ◐ *pressure ratio*	**relación de compresión**	taux de compression	2.3.2-**3**
compression stage	**etapa de compresión**	étage de compression	2.3.2-**4**
compression system *compression-type refrigerating system*	**sistema de compresión** instalación frigorífica de compresión	système frigorifique à compression *système à compression* ◐	2.3.2-**5**
compression-type refrigerating system *compression system*	**sistema de compresión** instalación frigorífica de compresión	système frigorifique à compression *système à compression* ◐	2.3.2-**6**
condensing pressure	**presión de condensación**	pression de condensation *pression de liquéfaction* ○	2.3.2-**7**
condensing temperature	**temperatura de condensación**	température de condensation	2.3.2-**8**
delivery pressure ○ *discharge pressure*	**presión de impulsión** presión de descarga	pression de refoulement	2.3.2-**9**
delivery temperature ○ *discharge temperature*	**temperatura de impulsión** temperatura de descarga	température de refoulement	2.3.2-**10**
desuperheating	**des-sobrecalentamiento** enfriamiento del vapor sobrecalentado	désurchauffe *refroidissement avant condensation* ◐	2.3.2-**11**
discharge	**impulsión** descarga	refoulement	2.3.2-**12**
discharge pressure *delivery pressure* ○	**presión de impulsión** presión de descarga	pression de refoulement	2.3.2-**13**
discharge temperature *delivery temperature* ○	**temperatura de impulsión** temperatura de descarga	température de refoulement	2.3.2-**14**
dry compression	**compresión seca** funcionamiento en régimen seco	fonctionnement en régime sec *fonctionnement en régime de surchauffe* ◐	2.3.2-**15**
dual compression ○	**compresión de doble aspiración**	compression à double aspiration ○	2.3.2-**16**
evaporating pressure	**presión de evaporación** presión de vaporización	pression d'évaporation *pression de vaporisation* ○	2.3.2-**17**
evaporating temperature	**temperatura de evaporación** temperatura de vaporización	température d'évaporation *température de vaporisation* ○	2.3.2-**18**
external coolant	**fluido para refrigerar el compresor**	fluide réfrigérant externe *fluide de refroidissement externe*	2.3.2-**19**
high-pressure side *high side* ◐	**lado de alta presión** alta presión	côté haute pression	2.3.2-**20**
high-pressure stage	**etapa de alta (presión)** compresión de alta	étage haute pression	2.3.2-**21**
high side ◐ *high-pressure side*	**lado de alta presión** alta presión	côté haute pression	2.3.2-**22**
inlet pressure	**presión de aspiración**	pression d'aspiration	2.3.2-**23**
inlet temperature	**temperatura de aspiración**	température d'aspiration	2.3.2-**24**
integrated cascade ◐ *mixed refrigerant cascade* ◐ *autorefrigerated cascade* ○	**cascada integrada**	cascade intégrée *cascade incorporée* ◐	2.3.2-**25**

	ENGLISH	ESPAÑOL	FRANÇAIS
2.3.2-**26**	intercooling *interstage cooling* ◐	**enfriamiento intermedio**	refroidissement intermédiaire
2.3.2-**27**	intermediate pressure *interstage pressure* ◐	**presión intermedia**	pression intermédiaire
2.3.2-**28**	interstage cooling ◐ *intercooling*	**enfriamiento intermedio**	refroidissement intermédiaire
2.3.2-**29**	interstage pressure ◐ *intermediate pressure*	**presión intermedia**	pression intermédiaire
2.3.2-**30**	low-pressure side *low side* ◐	**lado de baja presión** baja presión	côté basse pression
2.3.2-**31**	low-pressure stage	**etapa de baja (presión)**	étage basse pression
2.3.2-**32**	low side ◐ *low-pressure side*	**lado de baja presión** baja presión	côté basse pression
2.3.2-**33**	mixed refrigerant cascade ◐ *integrated cascade* ◐ *autorefrigerated cascade* ○	**cascada integrada**	cascade intégrée *cascade incorporée* ◐
2.3.2-**34**	multistage compression	**compresión en varias etapas** compresión múltiple	compression multiétagée *compression à plusieurs étages* ◐
2.3.2-**35**	multistage expansion	**expansión múltiple** expansión por etapas	détente fractionnée *détente étagée* ◐
2.3.2-**36**	pressure ratio *compression ratio* ◐	**relación de compresión**	taux de compression
2.3.2-**37**	shaft rotational speed	**revoluciones del eje del compresor**	fréquence de rotation (de l'arbre) *vitesse de rotation (de l'arbre)* ◐
2.3.2-**38**	single-stage compression	**compresión en una etapa** compresión simple	compression monoétagée *compression à un étage* ◐
2.3.2-**39**	stage pressure ratio	**relación de compresión de la etapa**	taux de compression par étage *rapport de pression par étage* ◐
2.3.2-**40**	suction	**aspiración**	aspiration
2.3.2-**41**	suction pressure	**presión de aspiración**	pression d'aspiration
2.3.2-**42**	suction temperature	**temperatura de aspiración**	température d'aspiration
2.3.2-**43**	total pressure ratio	**relación de compresión** tasa de compresión	taux de compression total *rapport total de compression* ◐
2.3.2-**44**	two-stage compression *compound compression*	**compresión compuesta** compresión en dos etapas compresión en doble salto	compression biétagée *compression à deux étages* ◐
2.3.2-**45**	volume factor	**factor de volumen**	coefficient de volume *coefficient de débit* ◐
2.3.2-**46**	wet compression	**compresión húmeda** funcionamiento en régimen húmedo	fonctionnement en régime humide *marche en régime humide* ◐ *fonctionnement humide* ◐

SECTION 2.3.3 *Other refrigerating systems*	**SUBCAPÍTULO 2.3.3** *Otros sistemas frigoríficos*	**SOUS-CHAPITRE 2.3.3** *Autres systèmes frigorifiques*
SECTION 2.3.3.1 *Sorption system*	**SUBCAPÍTULO 2.3.3.1** *Sistemas de sorción*	**SOUS-CHAPITRE 2.3.3.1** *Système à sorption*

	ENGLISH	ESPAÑOL	FRANÇAIS
2.3.3.1-**1**	absorbate	**absorbato**	absorbat
2.3.3.1-**2**	absorbent	**absorbente** (subst.)	absorbant

ENGLISH	ESPAÑOL	FRANÇAIS	
absorber	**absorbedor** (subst.)	absorbeur	2.3.3.1-**3**
absorption	**absorción**	absorption	2.3.3.1-**4**
absorption (refrigerating) machine	**máquina (frigorífica) de absorción**	machine (frigorifique) à absorption	2.3.3.1-**5**
absorption refrigerating system	**sistema (frigorífico) de absorción** instalación frigorífica de absorción	système frigorifique à absorption	2.3.3.1-**6**
absorption refrigerating unit	**grupo frigorífico de absorción**	groupe frigorifique à absorption	2.3.3.1-**7**
adsorbate	**adsorbato**	adsorbat	2.3.3.1-**8**
adsorbent	**adsorbente** (subst.)	adsorbant	2.3.3.1-**9**
adsorption	**adsorción**	adsorption	2.3.3.1-**10**
analyser	**deflegmador**	déphlegmateur	2.3.3.1-**11**
boiler ○ *generator* *regenerator* *desorber* ◑	**hervidor** (subst.) generador (subst.)	générateur *bouilleur* ○	2.3.3.1-**12**
desorber ◑ *generator* *regenerator* *boiler* ○	**hervidor** (subst.) generador (subst.)	générateur *bouilleur* ○	2.3.3.1-**13**
desorption	**desorción**	désorption	2.3.3.1-**14**
diffusion-absorption system	**sistema de absorción-difusión**	système à (absorption-)diffusion	2.3.3.1-**15**
double-effect cycle	**ciclo de doble efecto**	cycle à double effet	2.3.3.1-**16**
double-lift cycle	**ciclo de doble etapa**	cycle biétagé	2.3.3.1-**17**
exhaust steam	**vapor de escape**	vapeur d'échappement	2.3.3.1-**18**
generator *regenerator* *desorber* ◑ *boiler* ○	**hervidor** (subst.) generador (subst.)	générateur *bouilleur* ○	2.3.3.1-**19**
intermittent absorption refrigerating machine	**máquina frigorífica de absorción discontinua** máquina frigorífica de absorción intermitente	machine frigorifique à sorption à fonctionnement intermittent *machine frigorifique à absorption discontinue*	2.3.3.1-**20**
liquor ○ *solution*	**solución**	solution	2.3.3.1-**21**
multi-effect cycle	**ciclo multiefecto** ciclo de varias etapas	cycle multi-effets	2.3.3.1-**22**
multi-lift cycle	**ciclo multiefecto**	cycle multiétagé	2.3.3.1-**23**
multistage cycle	**ciclo multietapa** ciclo de varias etapas	cycle multiétagé	2.3.3.1-**24**
poor solution ◑ *weak solution*	**solución pobre**	solution faible *solution pauvre* *liqueur pauvre* ○	2.3.3.1-**25**
rectifier	**rectificador**	colonne de rectification *rectificateur* ◑	2.3.3.1-**26**
regenerator *generator* *desorber* ◑ *boiler* ○	**hervidor** (subst.) generador (subst.)	générateur *bouilleur* ○	2.3.3.1-**27**
resorption system	**sistema de resorción**	système à résorption	2.3.3.1-**28**

	ENGLISH	ESPAÑOL	FRANÇAIS
2.3.3.1-**29**	rich solution ◑ *strong solution*	**solución rica**	solution riche *solution forte* *liqueur riche* ○
2.3.3.1-**30**	single-effect cycle	**ciclo de una etapa**	cycle à simple effet
2.3.3.1-**31**	solution *liquor* ○	**solución**	solution
2.3.3.1-**32**	solution width	**diferencia de concentración de la solución**	écart de concentration *saut de titre* ◑
2.3.3.1-**33**	sorbate	**sorbato**	sorbat
2.3.3.1-**34**	sorbent	**sorbente**	sorbant
2.3.3.1-**35**	sorption	**sorción**	sorption
2.3.3.1-**36**	stripping column	**columna de reducción (de concentración de volátiles)**	colonne d'épuisement
2.3.3.1-**37**	strong solution *rich solution* ◑	**solución rica**	solution riche *solution forte* *liqueur riche* ○
2.3.3.1-**38**	weak solution *poor solution* ◑	**solución pobre**	solution faible *solution pauvre* *liqueur pauvre* ○

	SECTION 2.3.3.2 *Steam-jet system*	SUBCAPÍTULO 2.3.3.2 *Sistemas de eyección de vapor*	SOUS-CHAPITRE 2.3.3.2 *Système à éjection de vapeur*
2.3.3.2-**1**	ejector	**eyector** (subst.)	ejecto-compresseur
2.3.3.2-**2**	ejector-cycle refrigerating system ◑ *steam-jet refrigerating system*	**sistema frigorífico de eyección (de vapor)**	système frigorifique à éjection (de vapeur d'eau)
2.3.3.2-**3**	steam-jet refrigerating machine	**máquina frigorífica de eyección (de vapor)**	machine frigorifique à éjection (de vapeur d'eau)
2.3.3.2-**4**	steam-jet refrigerating system *ejector-cycle refrigerating system* ◑	**sistema frigorífico de eyección (de vapor)**	système frigorifique à éjection (de vapeur d'eau)

	SECTION 2.3.3.3 *Thermoelectric cooling*	SUBCAPÍTULO 2.3.3.3 *Enfriamiento termoeléctrico*	SOUS-CHAPITRE 2.3.3.3 *Refroidissement thermoélectrique*
2.3.3.3-**1**	figure of merit ◑	**coeficiente de mérito** grado de calidad	coefficient de mérite *facteur de mérite*
2.3.3.3-**2**	Peltier effect	**efecto Peltier**	effet Peltier
2.3.3.3-**3**	Seebeck effect	**efecto Seebeck**	effet Seebeck
2.3.3.3-**4**	semiconductor	**semiconductor**	semi-conducteur
2.3.3.3-**5**	thermoelectric battery ○ *thermoelectric module*	**módulo termoeléctrico**	module thermoélectrique
2.3.3.3-**6**	thermoelectric cooling *thermoelectric refrigeration* ◑	**enfriamiento termoeléctrico**	refroidissement thermoélectrique
2.3.3.3-**7**	thermoelectric module *thermoelectric battery* ○	**módulo termoeléctrico**	module thermoélectrique

ENGLISH	ESPAÑOL	FRANÇAIS	
thermoelectric refrigeration ◐ *thermoelectric cooling*	**enfriamiento termoeléctrico**	refroidissement thermoélectrique	2.3.3.3-**8**
thermopile	**termopila** pila termoeléctrica	pile thermoélectrique *thermopile*	2.3.3.3-**9**

SECTION 2.3.3.4 *Other refrigerating systems*	SUBCAPÍTULO 2.3.3.4 *Otros sistemas frigoríficos*	SOUS-CHAPITRE 2.3.3.4 *Autres systèmes frigorifiques*	
air-cycle refrigerating machine	**máquina (frigorífica) de ciclo de aire**	machine frigorifique à air	2.3.3.4-**1**
air (refrigeration) cycle	**ciclo (frigorífico) de aire**	cycle (frigorifique) à air	2.3.3.4-**2**
heat pipe	**tubo de calor** caloducto	caloduc	2.3.3.4-**3**
pulse-tube ○	**tubo pulsátil** tubo de pulsión	tube à pulsation ○	2.3.3.4-**4**
Ranque-Hilsch effect	**efecto Ranque-Hilsch**	effet Ranque-Hilsch	2.3.3.4-**5**
vortex tube	**tubo vórtex**	tube vortex *tube à tourbillon ◐*	2.3.3.4-**6**

SECTION 2.4 *Compressors* SECTION 2.4.1 *Compression refrigerating units*	SUBCAPÍTULO 2.4 *Compresores* SUBCAPÍTULO 2.4.1 *Grupos frigoríficos de compresión*	SECTION 2.4 *Compresseurs* SECTION 2.4.1 *Groupes frigorifiques à compression*	
accessible hermetic motor compressor ○ *semi-hermetic compressor*	**compresor semihermético** motocompresor hermético accesible	(moto)compresseur hermétique accessible *(moto)compresseur semi-hermétique ◐*	2.4.1-**1**
air-cooled unit	**unidad enfriada por aire**	groupe de condensation à air	2.4.1-**2**
commercial condensing unit ◐	**grupo frigorífico comercial**	groupe frigorifique commercial	2.4.1-**3**
compressor unit ◐	**grupo compresor** grupo motocompresor	motocompresseur *groupe motocompresseur ◐*	2.4.1-**4**
condensing unit	**unidad de condensación**	unité de condensation *groupe compresseur-condenseur ◐*	2.4.1-**5**
hermetically sealed condensing unit ○	**grupo frigorífico hermético**	groupe frigorifique hermétique	2.4.1-**6**
industrial condensing unit ○	**grupo frigorífico industrial**	groupe de condensation industriel	2.4.1-**7**
liquid-cooled unit	**unidad enfriada por agua**	groupe frigorifique à condensation par un liquide	2.4.1-**8**
low-capacity condensing unit	**grupo frigorífico de pequeña potencia** grupo frigorífico fraccionario	groupe frigorifique de faible puissance ○	2.4.1-**9**
open-type compressor	**compresor abierto** compresor (de grupo) abierto	compresseur ouvert	2.4.1-**10**
plug-in unit *plug unit*	**grupo frigorífico en caballete** grupo individual	groupe frigorifique de paroi *groupe frigorifique à insérer ◐*	2.4.1-**11**
plug unit *plug-in unit*	**grupo frigorífico en caballete** grupo individual	groupe frigorifique de paroi *groupe frigorifique à insérer ◐*	2.4.1-**12**
refrigerating unit	**grupo frigorífico**	groupe frigorifique	2.4.1-**13**
semi-hermetic compressor *accessible hermetic motor compressor ○*	**compresor semihermético** motocompresor hermético accesible	(moto)compresseur hermétique accessible *(moto)compresseur semi-hermétique ◐*	2.4.1-**14**

	ENGLISH	ESPAÑOL	FRANÇAIS
	SECTION 2.4.2 *Positive-displacement compressors*	**SUBCAPÍTULO 2.4.2** *Compresores de desplazamiento positivo*	**SOUS-CHAPITRE 2.4.2** *Compresseurs volumétriques*
2.4.2-1	booster compressor *low-stage compressor* ○	**compresor booster** compresor de retroalimentación	compresseur de suralimentation *compresseur basse pression* ○
2.4.2-2	closed crankcase compressor ○ *enclosed compressor* ○	**compresor de cárter cerrado**	compresseur (à carter) fermé
2.4.2-3	compound compressor ○	**compresor de doble salto** compresor de doble escalonamiento	compresseur biétagé *compresseur compound* ○
2.4.2-4	compressor	**compresor**	compresseur
2.4.2-5	diaphragm compressor	**compresor de membrana**	compresseur à membrane
2.4.2-6	differential piston compressor ○ *stepped piston compound compressor* ○	**compresor de pistón diferencial**	compresseur à pistons différentiels *compresseur à pistons étagés* ○
2.4.2-7	double-acting compressor	**compresor de doble efecto**	compresseur à double effet
2.4.2-8	dry piston compressor	**compresor de pistón seco** compresor de pistón de laberinto	compresseur sec *compresseur à piston sec* ○
2.4.2-9	dual compressor ○ *tandem compressor* ○	**compresores en tándem**	compresseurs jumelés *compresseurs en tandem* ○
2.4.2-10	dual-effect compressor	**compresor de doble aspiración**	compresseur avec orifice de suralimentation
2.4.2-11	enclosed compressor ○ *closed crankcase compressor* ○	**compresor de cárter cerrado**	compresseur (à carter) fermé
2.4.2-12	liquid-injected rotary compressor	**compresor rotativo con inyección de líquido**	compresseur rotatif à injection de fluide
2.4.2-13	low-stage compressor ○ *booster compressor*	**compresor booster** compresor de retroalimentación	compresseur de suralimentation *compresseur basse pression* ○
2.4.2-14	monoscrew compressor *single-screw compressor*	**compresor de tornillo de motor único**	compresseur à vis monomoteur
2.4.2-15	multistage compressor	**compresor multietapa** compresor de etapas múltiples	compresseur multiétagé
2.4.2-16	multivane rotary compressor	**compresor rotativo de paletas múltiples** compresor rotativo de paletas	compresseur rotatif à palettes multiples *compresseur rotatif multicellulaire* ○
2.4.2-17	oil-free compressor	**compresor sin aceite**	compresseur non lubrifié
2.4.2-18	open crankcase compressor ○	**compresor de cárter abierto**	compresseur (à carter) ouvert ○
2.4.2-19	open-type compressor	**compresor abierto** compresor (de grupo) abierto	compresseur ouvert
2.4.2-20	positive-displacement compressor ○	**compresor volumétrico** compresor de desplazamiento positivo	compresseur volumétrique
2.4.2-21	reciprocating compressor	**compresor alternativo** compresor de émbolo compresor de pistón	compresseur alternatif *compresseur à piston*
2.4.2-22	refrigerant compressor	**compresor frigorífico**	compresseur frigorifique
2.4.2-23	return-flow compressor ○	**compresor de flujo alterno**	compresseur à flux alternatif
2.4.2-24	rolling piston compressor	**compresor de pistón rotativo**	compresseur à piston tournant *compresseur à piston rotatif* ○
2.4.2-25	rotary compressor	**compresor rotativo**	compresseur rotatif
2.4.2-26	screw compressor	**compresor de tornillo**	compresseur à vis *compresseur hélicoïde* ○

ENGLISH	ESPAÑOL	FRANÇAIS	
scroll compressor	**compresor espiro-orbital**	compresseur à spirale *compresseur spiro-orbital* ◐	2.4.2-**27**
single acting compressor	**compresor de simple efecto**	compresseur à simple effet	2.4.2-**28**
single-screw compressor *monoscrew compressor*	**compresor de tornillo de motor único**	compresseur à vis monomoteur	2.4.2-**29**
single-vane rotary compressor ◐	**compresor rotativo de paleta**	compresseur rotatif à palette unique *compresseur rotatif monocellulaire* ○	2.4.2-**30**
sliding-vane compressor ◐	**compresor de paletas**	compresseur à palettes	2.4.2-**31**
standard discharge point	**punto estándar de descarga**	orifice de refoulement *refoulement*	2.4.2-**32**
standard inlet point	**punto estándar de aspiración** punto estándar de admisión	orifice d'aspiration *aspiration*	2.4.2-**33**
stepped piston compound compressor ○ *differential piston compressor* ◐	**compresor de pistón diferencial**	compresseur à pistons différentiels *compresseur à pistons étagés* ◐	2.4.2-**34**
swash plate compressor	**compresor de placa oscilante**	compresseur volumétrique à barillet	2.4.2-**35**
tandem compressor ◐ *dual compressor* ◐	**compresores en tándem**	compresseurs jumelés *compresseurs en tandem* ◐	2.4.2-**36**
twin cylinder compressor	**compresor de dos cilindros**	compresseur à deux cylindres	2.4.2-**37**
uniflow compressor ○	**compresor de flujo continuo** compresor en equicorriente	compresseur à flux continu *compresseur à équicourant* ◐	2.4.2-**38**
variable displacement wobble plate compressor	**compresor volumétrico de plato oscilante**	compresseur volumétrique variable à plateau oscillant	2.4.2-**39**

SECTION 2.4.3 *Components of positive-displacement compressors*	SUBCAPÍTULO 2.4.3 *Elementos de los compresores de desplazamiento positivo*	SOUS-CHAPITRE 2.4.3 *Eléments des compresseurs volumétriques*	
actual displacement	**volumen desplazado real**	débit volume aspiré (m³/h) *débit volume déplacé réel* ◐	2.4.3-**1**
ball bearing	**cojinete de bolas**	palier à billes *roulement à billes* ◐	2.4.3-**2**
beam valve *leaf valve*	**válvula de lámina**	soupape à lamelle(s)	2.4.3-**3**
bearing	**soporte de cojinete** cojinete	palier *coussinet*	2.4.3-**4**
bearing housing	**caja de cojinete**	boîtier de palier *cage de palier* ◐	2.4.3-**5**
bellows seal	**cierre de cigüeñal de fuelle**	garniture d'étanchéité à soufflet	2.4.3-**6**
big end	**cabeza de biela**	tête de bielle	2.4.3-**7**
cantilever valve ◐ *flapper valve*	**válvula de lámina**	soupape à lamelles multiples *soupape à languette* ◐ *clapet fléchissant* ◐ *clapet battant* ◐	2.4.3-**8**
clearance pocket ○	**espacio perjudicial adicional** espacio muerto adicional	espace mort additionnel *poche d'espace mort* ○	2.4.3-**9**
clearance volume	**espacio perjudicial** espacio muerto	espace mort	2.4.3-**10**

	ENGLISH	ESPAÑOL	FRANÇAIS
2.4.3-**11**	clearance volume ratio	**espacio perjudicial relativo** espacio muerto relativo	pourcentage d'espace mort *taux d'espace mort* ◑
2.4.3-**12**	compression stroke	**carrera de compresión**	course de compression *compression (phase de)*
2.4.3-**13**	(compressor) valve	**válvula**	soupape *clapet* ◑
2.4.3-**14**	connecting rod	**biela**	bielle
2.4.3-**15**	coupling (of shafts)	**acoplamiento (de ejes)**	accouplement direct
2.4.3-**16**	crankcase	**cárter**	carter
2.4.3-**17**	crankcase seal *(shaft) seal*	**caja de cierre** cierre de cigüeñal cárter	garniture d'étanchéité
2.4.3-**18**	crankpin	**codo de cigüeñal** mangueta de cigüeñal	maneton
2.4.3-**19**	crankshaft	**cigüeñal**	vilebrequin
2.4.3-**20**	cylinder	**cilindro**	cylindre
2.4.3-**21**	cylinder block	**bloque de cilindros**	bloc cylindre
2.4.3-**22**	(cylinder) bore	**diámetro interior (del cilindro)**	alésage
2.4.3-**23**	cylinder head	**culata del cilindro** cabeza del cilindro	culasse *tête de cylindre* ◑
2.4.3-**24**	cylinder liner	**camisa desmontable del cilindro**	chemise de cylindre amovible
2.4.3-**25**	cylinder wall	**pared interior del cilindro**	paroi (intérieure) du cylindre *surface intérieure du cylindre*
2.4.3-**26**	delivery valve ◑ *discharge valve* *outlet valve* ○	**válvula de impulsión** válvula de descarga	soupape de refoulement
2.4.3-**27**	diaphragm valve ◑ *disc valve*	**válvula de disco**	clapet à disque
2.4.3-**28**	disc valve *diaphragm valve* ◑	**válvula de disco**	clapet à disque
2.4.3-**29**	discharge line valve ◑ *discharge stop valve* ◑	**válvula de cierre de impulsión**	robinet de refoulement *robinet d'arrêt au refoulement* ◑
2.4.3-**30**	discharge stop valve ◑ *discharge line valve* ◑	**válvula de cierre de impulsión**	robinet de refoulement *robinet d'arrêt au refoulement* ◑
2.4.3-**31**	discharge stroke ◑	**carrera de impulsión**	course de refoulement
2.4.3-**32**	discharge valve *delivery valve* ◑ *outlet valve* ○	**válvula de impulsión** válvula de descarga	soupape de refoulement
2.4.3-**33**	displacement	**desplazamiento**	volume balayé *débit volumétrique* ◑
2.4.3-**34**	eccentric strap	**collar de excéntrica** biela de excéntrica	collier d'excentrique
2.4.3-**35**	economizer connexion *side port* ◑	**orificio de sobrealimentación**	orifice de suralimentation *orifice économiseur* ◑
2.4.3-**36**	expansion stroke	**carrera de expansión**	course de détente
2.4.3-**37**	flapper valve *cantilever valve* ◑	**válvula de lámina**	soupape à lamelles multiples *soupape à languette* ◑ *clapet fléchissant* ◑ *clapet battant* ◑

ENGLISH	ESPAÑOL	FRANÇAIS	
flexing valve o *reed valve* o	**válvula de lengüeta**	soupape à lamelles multiples *soupape flexible* o *soupape à ruban* o	2.4.3-**38**
footstep bearing	**quicionera** rangua	palier de butée vertical *crapaudine* o	2.4.3-**39**
friction ring o *rubbing ring*	**anillo de estanqueidad**	bague d'étanchéité *anneau d'étanchéité* o	2.4.3-**40**
gland packing o *shaft packing* o	**empaquetadura de estanqueidad** guarnición de estanqueidad	tresse graphitée d'étanchéité *garniture d'étanchéité*	2.4.3-**41**
gudgeon pin o *piston pin* *wrist pin* o	**bulón de émbolo** bulón de pistón	axe de piston *axe de pied de bielle* o	2.4.3-**42**
(guided) vane	**paleta deslizante**	palette libre	2.4.3-**43**
hydraulic thrust	**pistón de equilibrado**	piston d'équilibrage	2.4.3-**44**
journal	**gorrón**	tourillon	2.4.3-**45**
labyrinth seal	**cierre de laberinto**	labyrinthe	2.4.3-**46**
leaf valve *beam valve*	**válvula de lámina**	soupape à lamelle(s)	2.4.3-**47**
little end *small end* o	**pie de biela**	pied de bielle	2.4.3-**48**
(mechanical) clearance	**juego (mecánico)**	jeu (mécanique)	2.4.3-**49**
mechanical seal	**cierre de cigüeñal mecánico**	garniture mécanique	2.4.3-**50**
mushroom valve	**válvula de asiento cónico**	soupape tronconique *soupape champignon* o	2.4.3-**51**
needle bearing	**cojinete de agujas**	palier à aiguilles	2.4.3-**52**
outlet valve o *discharge valve* *delivery valve* o	**válvula de impulsión** válvula de descarga	soupape de refoulement	2.4.3-**53**
partial duty port	**lumbrera de reducción de potencia**	lumière de réduction de puissance	2.4.3-**54**
pedestal bearing	**soporte de cojinete auxiliar**	palier support auxiliaire	2.4.3-**55**
piston	**émbolo** pistón	piston	2.4.3-**56**
piston displacement o *swept volume* o	**volumen desplazado** volumen barrido	cylindrée	2.4.3-**57**
piston pin *gudgeon pin* o *wrist pin* o	**bulón de émbolo** bulón de pistón	axe de piston *axe de pied de bielle* o	2.4.3-**58**
piston ring	**segmento de émbolo** segmento de pistón	segment d'étanchéité	2.4.3-**59**
piston stroke	**carrera del émbolo** carrera del pistón	course	2.4.3-**60**
poppet valve	**válvula de vástago** válvula de asiento cónico	soupape à queue	2.4.3-**61**
reed valve o *flexing valve* o	**válvula de lengüeta**	soupape à lamelles multiples *soupape flexible* o *soupape à ruban* o	2.4.3-**62**
ring plate valve o *ring valve*	**válvula anular**	soupape annulaire *soupape concentrique*	2.4.3-**63**
ring valve *ring plate valve* o	**válvula anular**	soupape annulaire *soupape concentrique*	2.4.3-**64**

	ENGLISH	ESPAÑOL	FRANÇAIS
2.4.3-65	roller bearing	**cojinete de rodillos**	palier à rouleaux
2.4.3-66	rotary seal	**cierre rotativo**	garniture d'étanchéité mécanique
2.4.3-67	rubbing ring *friction ring* ◑	**anillo de estanqueidad**	bague d'étanchéité *anneau d'étanchéité* ◑
2.4.3-68	safety (cylinder) head	**cabeza de cilindro de seguridad** culata de cilindro de seguridad	fond mobile (sécurité anti-coup de liquide) *tête de cylindre à sécurité* ◑
2.4.3-69	safety (valve) head ◑	**asiento de válvula de seguridad** cabeza de seguridad	soupape de refoulement de sécurité ◑
2.4.3-70	scraper ring	**segmento rascador de aceite** segmento arrastrador de aceite	segment racleur
2.4.3-71	shaft packing ◑ *gland packing* ◑	**empaquetadura de estanqueidad** guarnición de estanqueidad	tresse graphitée d'étanchéité *garniture d'étanchéité*
2.4.3-72	(shaft) seal *crankcase seal*	**caja de cierre** cierre de cigüeñal cárter	garniture d'étanchéité
2.4.3-73	side port ◑ *economizer connexion*	**orificio de sobrealimentación**	orifice de suralimentation *orifice économiseur* ◑
2.4.3-74	sleeve bearing	**cojinete plano**	palier lisse
2.4.3-75	slide valve	**tirador de variación de potencia**	tiroir de variation de puissance *tiroir de variation de volume*
2.4.3-76	small end ◑ *little end*	**pie de biela**	pied de bielle
2.4.3-77	stuffing box	**prensaestopas** caja de estopadas	presse-étoupe (garniture de)
2.4.3-78	suction inlet	**orificio de aspiración**	orifice d'aspiration
2.4.3-79	suction line valve *suction stop valve* ◑	**válvula de cierre de aspiración**	robinet d'aspiration
2.4.3-80	suction stop valve ◑ *suction line valve*	**válvula de cierre de aspiración**	robinet d'aspiration
2.4.3-81	suction stroke	**carrera de aspiración**	course d'aspiration
2.4.3-82	suction valve	**válvula de aspiración**	clapet d'aspiration *soupape d'aspiration* ◑
2.4.3-83	swept volume ◑ *piston displacement* ◑	**volumen desplazado** volumen barrido	cylindrée
2.4.3-84	theoretical displacement (1)	**volumen desplazado teórico**	volume engendré
2.4.3-85	theoretical displacement (2)	**volumen desplazado teórico**	volume balayé *volume engendré*
2.4.3-86	thrust bearing	**cojinete de empuje**	palier de butée
2.4.3-87	thrust collar	**anillo de empuje**	collet de butée
2.4.3-88	tulip valve	**válvula de vástago**	soupape tulipe
2.4.3-89	valve cage	**caja de válvula**	boîtier à soupape *boîte à soupape* *cage de soupape*
2.4.3-90	valve cover	**tapa de válvula**	couvercle de soupape *chapeau de soupape* ○
2.4.3-91	valve guard	**guardaválvula** limitador del salto	contre clapet *butée de soupape* ◑
2.4.3-92	valve lift	**salto de válvula**	levée *levée de clapet* ◑
2.4.3-93	valve plate	**plato de válvula**	plaque porte-clapet *plaque porte-soupape* ◑

ENGLISH	ESPAÑOL	FRANÇAIS	
valve port	**orificio de paso**	lumière de passage	2.4.3-**94**
valve seat	**asiento de válvula**	siège de soupape	2.4.3-**95**
valve stem	**vástago de válvula**	tige de soupape	2.4.3-**96**
Vi *volume index*	**relación de volúmen**	volume index *Vi*	2.4.3-**97**
volume index Vi	**relación de volúmen**	volume index *Vi*	2.4.3-**98**
water (cooled) jacket	**camisa de agua**	chemise d'eau *culasse refroidie par eau* ○	2.4.3-**99**
wrist pin ○ *piston pin* *gudgeon pin* ○	**bulón de émbolo** bulón de pistón	axe de piston *axe de pied de bielle* ○	2.4.3-**100**

SECTION 2.4.4 *Turbocompressors*	SUBCAPÍTULO 2.4.4 *Turbocompresores*	SOUS-CHAPITRE 2.4.4 *Turbocompresseurs*	
axial flow compressor	**compresor (de flujo) axial** turbocompresor	turbocompresseur axial	2.4.4-**1**
blade *vane* ○	**álabe**	ailette *aube* *pale* ○	2.4.4-**2**
blade flutter	**vibración del álabe**	vibration d'ailette *vibration de pale* ○	2.4.4-**3**
blade passing frequency	**frecuencia debida al paso de los álabes**	fréquence due aux sillages d'ailettes	2.4.4-**4**
casing (of a turbocompressor)	**carcasa (de un compresor centrífugo)** cuerpo (de un compresor centrífugo)	corps de compresseur	2.4.4-**5**
centrifugal compressor	**compresor centrífugo**	compresseur centrifuge	2.4.4-**6**
diffuser	**difusor**	diffuseur	2.4.4-**7**
drag coefficient	**coeficiente de arrastre**	trainée	2.4.4-**8**
eye ○ *inlet*	**oído (de entrada)**	ouïe (d'entrée)	2.4.4-**9**
flexible-shaft centrifugal compressor	**compresor centrífugo de árbol flexible**	compresseur centrifuge à arbre flexible *compresseur à arbre souple* ○	2.4.4-**10**
flow coefficient for a dynamic compressor	**coeficiente de flujo (del turbocompresor)**	invariant de débit d'un étage de turbo-compresseur	2.4.4-**11**
impeller *rotor* *wheel* ○	**rodete**	roue *rotor* *mobile* ○	2.4.4-**12**
impeller reaction	**grado de reacción**	degré de réaction (dans la roue) *taux de réaction (dans la roue)* ○	2.4.4-**13**
impeller running frequency	**frecuencia debida a la rotación del rodete**	fréquence sonore	2.4.4-**14**
inlet *eye* ○	**oído (de entrada)**	ouïe (d'entrée)	2.4.4-**15**
non-positive displacement compressor	**compresor cinético**	compresseur cinétique	2.4.4-**16**
open impeller *unshrouded impeller* ○	**rodete abierto**	roue ouverte	2.4.4-**17**
overall pressure coefficient for a dynamic compressor	**coeficiente de presión total de un compresor dinámico**	invariant de pression	2.4.4-**18**

	ENGLISH	ESPAÑOL	FRANÇAIS
2.4.4-**19**	pre-rotary vane ◐ *pre-rotation vane*	**álabe de prerrotación** guía de prerrotación	aube de prérotation *ventelle ◐*
2.4.4-**20**	pre-rotation vane *pre-rotary vane ◐*	**álabe de prerrotación** guía de prerrotación	aube de prérotation *ventelle ◐*
2.4.4-**21**	pre-rotary vane assembly	**mecanismo de álabes de prerrotación**	aubage (directeur) de prérotation
2.4.4-**22**	rotor *impeller* *wheel* ○	**rodete**	roue *rotor* *mobile ◐*
2.4.4-**23**	scroll *volute ◐*	**voluta**	volute
2.4.4-**24**	shrouded impeller	**rodete cerrado**	roue fermée
2.4.4-**25**	slip	**despegue**	décollement
2.4.4-**26**	stall (of a turbocompressor or centrifugal compressor) *stalling ◐*	**bloqueo (de un turbocompresor)**	blocage
2.4.4-**27**	stalling ◐ *stall (of a turbocompressor or centrifugal compressor)*	**bloqueo (de un turbocompresor)**	blocage
2.4.4-**28**	stiff-shaft centrifugal compressor	**compresor centrífugo de árbol rígido**	compresseur centrifuge à arbre rigide
2.4.4-**29**	subsonic compressor	**compresor subsónico**	compresseur subsonique
2.4.4-**30**	supersonic compressor	**compresor supersónico**	compresseur supersonique
2.4.4-**31**	surging (of a turbocompressor or centrifugal compressor)	**fluctuación** bombeo	pompage (d'une turbomachine)
2.4.4-**32**	swirl	**torbellino**	tourbillon
2.4.4-**33**	total to static efficiency ◐	**rendimiento isoentrópico** rendimiento estático	rendement isentropique
2.4.4-**34**	total to total efficiency ◐	**rendimiento politrópico** rendimiento total	rendement polytropique
2.4.4-**35**	turbocompressor	**turbocompresor**	turbocompresseur *turbomachine*
2.4.4-**36**	unshrouded impeller ◐ *open impeller*	**rodete abierto**	roue ouverte
2.4.4-**37**	vane ◐ *blade*	**álabe**	ailette *aube* *pale ◐*
2.4.4-**38**	volute ◐ *scroll*	**voluta**	volute
2.4.4-**39**	wheel ○ *impeller* *rotor*	**rodete**	roue *rotor* *mobile ◐*

SECTION 2.5 *Heat exchangers*	**SUBCAPÍTULO 2.5** *Cambiadores de calor*	**SOUS-CHAPITRE 2.5** *Echangeurs thermiques*
SECTION 2.5.1 *Heat exchangers:* *general background*	**SUBCAPÍTULO 2.5.1** *Generalidades sobre cambiadores* *de calor*	**SOUS-CHAPITRE 2.5.1** *Généralités sur les échangeurs* *thermiques*

	ENGLISH	ESPAÑOL	FRANÇAIS
2.5.1-**1**	baffle ◐ *baffle plate*	**placa deflectora** tabique deflector	chicane

ENGLISH	ESPAÑOL	FRANÇAIS	
baffle plate *baffle* ⚬	**placa deflectora** tabique deflector	chicane	2.5.1-**2**
baudelot cooler ⚬ *irrigation cooler* ⚬ *surface liquid cooler* ⚬	**enfriador de cortina**	refroidisseur à ruissellement	2.5.1-**3**
bottle-type liquid cooler	**enfriador de botella**	fontaine réfrigérée	2.5.1-**4**
brazed plate heat exchanger	**cambiador de calor de placas**	échangeur à plaques *échangeur à plaques brasées ou soudées*	2.5.1-**5**
brushed surface	**cambiador de superficie rascada** cambiador de rascador	surface brossée	2.5.1-**6**
bundle of tubes	**haz de tubos** haz tubular	faisceau tubulaire *faisceau de tubes* ⚬	2.5.1-**7**
coil depth *core depth*	**profundidad de aleta** profundidad del serpentín	profondeur ailetée (d'une batterie) *profondeur*	2.5.1-**8**
coil face area *core face area*	**superficie frontal** sección frontal	section frontale	2.5.1-**9**
coil length *core length*	**longitud aleteada**	longueur ailetée (d'une batterie)	2.5.1-**10**
coil width *core width*	**anchura aleteada**	largeur ailetée (d'une batterie)	2.5.1-**11**
compact heat exchanger	**cambiador de calor compacto**	échangeur (thermlque) compact	2.5.1-**12**
cooling surface	**superficie de enfriamiento**	surface froide	2.5.1-**13**
core depth *coil depth*	**profundidad de aleta** profundidad del serpentín	profondeur ailetée (d'une batterie) *profondeur*	2.5.1-**14**
core face area *coil face area*	**superficie frontal** sección frontal	section frontale	2.5.1-**15**
core length *coil length*	**longitud aleteada**	longueur ailetée (d'une batterie)	2.5.1-**16**
core width *coil width*	**anchura aleteada**	largeur ailetée (d'une batterie)	2.5.1-**17**
counterflow arrangement	**circulación a contracorriente** flujo a contracorriente	circulation à contre-courant	2.5.1-**18**
counterflow exchange	**intercambio de calor a contracorriente**	échange à contre-courant	2.5.1-**19**
counterflow heat exchanger	**cambiador de calor a contracorriente**	échangeur (thermique) à contrecourant	2.5.1-**20**
cross-flow exchange	**intercambio de calor de flujos cruzados**	échange à courants croisés	2.5.1-**21**
cross-flow heat exchanger	**cambiador de calor de flujos cruzadas**	échangeur (thermique) à courants croisés	2.5.1-**22**
desuperheater	**des-sobrecalentador** recuperador del sobrecalentamiento	désurchauffeur	2.5.1-**23**
direct-contact heat exchanger	**cambiador de calor mezcla** cambiador de calor de contacto directo	échangeur (thermique) à contact direct *échangeur par mélange*	2.5.1-**24**
direct surface ⚬ *primary surface* ⚬	**superficie primaria** superficie directa	surface primaire *surface directe*	2.5.1-**25**
drum cooler	**enfriador de tambor**	refroidisseur à tambour rotatif	2.5.1-**26**
dry cooler	**enfriador seco**	refroidisseur de fluide atmosphérique sec *aérorefroidisseur sec* ⚬	2.5.1-**27**
enclosed liquid cooler	**enfriador de líquidos con evaporación extratubular**	refroidisseur de liquide à évaporation extratubulaire	2.5.1-**28**
enhanced surface	**tubos de superficie aumentada** tubos de superficie acrecentada	tubes à surface améliorée	2.5.1-**29**

	ENGLISH	ESPAÑOL	FRANÇAIS
2.5.1-30	extended surface	**superficie adicional**	surface "augmentée" *surface "étendue"*
2.5.1-31	face area	**sección frontal** superficie frontal	section frontale
2.5.1-32	fin	**aleta**	ailette
2.5.1-33	fin efficiency	**efectividad de la aleta**	rendement d'ailette
2.5.1-34	fin pitch	**paso de las aletas**	nombre d'ailettes par unité de longueur
2.5.1-35	finned length	**longitud útil del tubo aleteado**	longueur ailetée *longueur utile d'un tube à ailettes*
2.5.1-36	finned surface	**superficie aletada** superficie extendida	surface ailetée *surface des ailettes ◐*
2.5.1-37	finned tube	**tubo de aletas**	tube à ailettes
2.5.1-38	flow area *free flow area*	**sección de paso**	section de passage *section claire ◐*
2.5.1-39	fouling	**incrustación**	encrassement
2.5.1-40	fouling resistance	**resistencia térmica de la incrustación**	résistance thermique d'encrassement
2.5.1-41	free flow area *flow area*	**sección de paso**	section de passage *section claire ◐*
2.5.1-42	gas cooler	**enfriador de gases**	refroidisseur de gaz
2.5.1-43	grid coil ○ *hairpin coil ◐*	**serpentín de horquilla**	échangeur en épingle à cheveux
2.5.1-44	hairpin coil ◐ *grid coil ○*	**serpentín de horquilla**	échangeur en épingle à cheveux
2.5.1-45	heat exchange surface	**superficie de intercambio (térmico)**	surface d'échange (thermique)
2.5.1-46	heat exchanger	**cambiador de calor**	échangeur thermique *échangeur de chaleur ◐*
2.5.1-47	heat recovery heat exchanger	**cambiador de calor con recuperación**	échangeur de récupération de chaleur
2.5.1-48	heat recovery liquid chilling package	**unidad autónoma de enfriamiento de líquidos con recuperación de calor**	groupe refroidisseur de liquide récupérateur de chaleur
2.5.1-49	heat transfer surface (air side)	**superficie de transmisión de calor (lado aire)**	surface d'échange de chaleur (côté air)
2.5.1-50	heating surface	**superficie de calentamiento** superficie de caldeo	surface de chauffe
2.5.1-51	helical fin ◐ *spiral fin ◐* *strip fin ◐*	**aleta helicoidal** aleta en espiral	ailette spiralée *ailette hélicoïdale ◐*
2.5.1-52	ice bank cooler	**enfriador de agua con acumulación de hielo**	refroidisseur d'eau accumulateur de glace
2.5.1-53	in-line bank of tubes	**fila de tubos en linea**	nappe de tubes alignés
2.5.1-54	indirect surface ◐ *secondary surface*	**superficie secundaria** superficie indirecta	surface secondaire *surface indirecte*
2.5.1-55	inlet temperature	**temperatura de entrada** temperatura de admisión	température d'entrée
2.5.1-56	inner fin	**aleta interior**	ailette interne *ailette intérieure ◐*
2.5.1-57	insert ◐ *turbulator*	**generador de turbulencia**	turbulateur *insert ◐*
2.5.1-58	integral fin	**aleta integral**	ailette extrudée
2.5.1-59	internal volume	**volumen interno**	volume interne

ENGLISH	ESPAÑOL	FRANÇAIS	
irrigation cooler ◐ *surface liquid cooler* ◐ *baudelot cooler* ○	**enfriador de cortina**	refroidisseur à ruissellement	2.5.1-**60**
liquid cooler	**enfriador de líquidos**	refroidisseur de liquide	2.5.1-**61**
liquid inlet temperature	**temperatura de entrada (de líquido)** temperatura de admisión (de líquido)	température du liquide à l'entrée	2.5.1-**62**
liquid outlet temperature	**temperatura de salida (de líquido)** temperatura de retorno (de líquido)	température du liquide à la sortie	2.5.1-**63**
longitudinal fin	**aleta longitudinal**	ailette longitudinale	2.5.1-**64**
method of heat transfer	**sistema (método) de transmisión de calor**	mode d'échange thermique	2.5.1-**65**
microchannel heat exchanger	**cambiador de calor de microcanales**	échangeur (thermique) à micro canaux	2.5.1-**66**
multipipe heat exchanger ◐ *multitubular heat exchanger*	**cambiador de calor multitubular**	échangeur (thermique) multitubulaire	2.5.1-**67**
multichannel heat exchanger	**cambiador de calor de canales múltiples**	échangeur (thermique) multicanaux	2.5.1-**68**
multitubular heat exchanger *multipipe heat exchanger* ◐	**cambiador de calor multitubular**	échangeur (thermique) multitubulaire	2.5.1-**69**
number of passes	**número de pasos**	nombre de passes	2.5.1-**70**
outlet temperature	**temperatura de salida** temperatura de retorno	température de sortie	2.5.1-**71**
parallel flow arrangement	**circulación a equicorriente** flujo a equicorriente	circulation à cocourant *circulation à équicourant* ◐	2.5.1-**72**
parallel flow heat exchanger	**cambiador de calor a equicorriente**	échangeur (thermique) à cocourant *échangeur (thermique) à équicourant* ◐	2.5.1-**73**
(pipe) coil	**serpentín**	serpentin	2.5.1-**74**
plain tube *smooth tube* *smooth pipe* ◐	**tubo liso**	tube lisse	2.5.1-**75**
plate fin	**aleta de placa**	ailette multiple	2.5.1-**76**
plate heat exchanger	**cambiador de calor de placas**	échangeur (thermique) à plaques	2.5.1-**77**
plate liquid cooler	**enfriador de placas**	refroidisseur de liquide à plaques	2.5.1-**78**
primary surface ◐ *direct surface* ○	**superficie primaria** superficie directa	surface primaire *surface directe*	2.5.1-**79**
removable end cover *removable head* ○	**fondo desmontable**	fond démontable *boîte d'extrémité démontable* ◐	2.5.1-**80**
removable head ○ *removable end cover*	**fondo desmontable**	fond démontable *boîte d'extrémité démontable* ◐	2.5.1-**81**
rotary heat exchanger	**cambiador de calor rotativo**	échangeur thermique rotatif	2.5.1-**82**
row	**fila**	rangée	2.5.1-**83**
row of tubes ◐	**fila de tubos**	nappe de tubes *rideau de tubes*	2.5.1-**84**
scraped heat exchanger	**cambiador de calor de superficie rascada** cambiador de calor con equipo exterior raspador	échangeur thermique à surface raclée	2.5.1-**85**
secondary surface *indirect surface* ◐	**superficie secundaria** superficie indirecta	surface secondaire *surface indirecte*	2.5.1-**86**
shell-and-tube heat exchanger	**cambiador de tubos y envolvente** cambiador multitubular de envolvente	échangeur (thermique) à (tubes et) calandre	2.5.1-**87**

	ENGLISH	ESPAÑOL	FRANÇAIS
2.5.1-**88**	smooth pipe ◐ *plain tube* *smooth tube*	**tubo liso**	tube lisse
2.5.1-**89**	smooth tube *plain tube* *smooth pipe* ◐	**tubo liso**	tube lisse
2.5.1-**90**	spiral fin ◐ *helical fin* ◐ *strip fin* ◐	**aleta helicoidal** aleta en espiral	ailette spiralée *ailette hélicoïdale* ◐
2.5.1-**91**	staggered bank of tubes	**fila de tubos al tresbolillo**	nappes de tubes en quinconce
2.5.1-**92**	strip fin ◐ *helical fin* ◐ *spiral fin* ◐	**aleta helicoidal** aleta en espiral	ailette spiralée *ailette hélicoïdale* ◐
2.5.1-**93**	surface liquid cooler ◐ *irrigation cooler* ◐ *baudelot cooler* ○	**enfriador de cortina**	refroidisseur à ruissellement
2.5.1-**94**	(thermal) recuperator *(thermal) regenerator*	**regenerador (térmico)** recuperador (térmico)	récupérateur thermique *régénérateur thermique* ◐
2.5.1-**95**	(thermal) regenerator *(thermal) recuperator*	**regenerador (térmico)** recuperador (térmico)	récupérateur thermique *régénérateur thermique* ◐
2.5.1-**96**	tube plate ◐ *tube sheet*	**placa tubular**	plaque tubulaire
2.5.1-**97**	tube sheet *tube plate* ◐	**placa tubular**	plaque tubulaire
2.5.1-**98**	turbulator *insert* ◐	**generador de turbulencia**	turbulateur *insert* ◐
2.5.1-**99**	water chiller *water cooler* ◐	**enfriador de agua**	refroidisseur d'eau
2.5.1-**100**	water cooler ◐ *water chiller*	**enfriador de agua**	refroidisseur d'eau
2.5.1-**101**	welded plate heat exchanger	**cambiador de calor de placas soldadas**	échangeur à cassettes

	SECTION 2.5.2 *Condensers and condenser accessories*	SUBCAPÍTULO 2.5.2 *Condensadores y sus accesorios*	SOUS-CHAPITRE 2.5.2 *Condenseurs et accessoires*
2.5.2-**1**	air-cooled condenser	**condensador enfriado por aire**	condenseur à air *aérocondenseur* ◐
2.5.2-**2**	algaecide	**algicida**	algicide
2.5.2-**3**	approach	**diferencial** aproximación	approche
2.5.2-**4**	atmospheric condenser	**condensador atmosférico** condensador de lluvia	condenseur à ruissellement
2.5.2-**5**	atmospheric cooling tower ○ *natural-draught cooling tower*	**torre de enfriamiento de circulación natural** torre de enfriamiento de agua con tiro natural	refroidisseur d'eau atmosphérique à tirage nature *tour de refroidissement à tirage naturel*
2.5.2-**6**	back-up valve	**válvula limitadora de presión** válvula de contrapresión	robinet à contre pression
2.5.2-**7**	barometric condenser	**condensador barométrico**	condenseur barométrique
2.5.2-**8**	bleeder-type condenser	**condensador de extracción**	condenseur à soutirage

ENGLISH	ESPAÑOL	FRANÇAIS	
blow-down water	**agua de descarga**	purge de déconcentration	2.5.2-**9**
circulating water	**agua en circulación**	eau de circulation	2.5.2-**10**
closed shell-and-tube condenser	**condensador multitubular de envolvente cerrado**	condenseur à calandre fermé	2.5.2-**11**
cold water	**agua fria**	eau refroidie	2.5.2-**12**
cold water basin	**depósito de agua fría**	bassin d'eau refroidie	2.5.2-**13**
condenser	**condensador**	condenseur	2.5.2-**14**
condenser-evaporator	**condensador-evaporador**	condenseur évaporateur	2.5.2-**15**
condenser-receiver	**condensador-recipiente**	condenseur-réservoir *condenseur accumulateur de liquide* o	2.5.2-**16**
condenser subcooling	**subenfriamiento en el condensador**	sous- refroidissement (en sortie de condenseur)	2.5.2-**17**
cooling tower	**torre de enfriamiento**	refroidisseur d'eau atmosphérique *tour de refroidissement (d'eau)*	2.5.2-**18**
desuperheating coil	**serpentín preenfriador**	échangeur désurchauffeur *élément désurchauffeur* o	2.5.2-**19**
dry cooling tower	**torre de enfriamiento seca**	tour de refroidissement sèche *refroidisseur d'eau atmosphérique sec*	2.5.2-**20**
evaporative condenser	**condensador evaporativo**	condenseur évaporatif *condenseur à évaporation d'eau* o *condenseur évaporateur d'eau* o	2.5.2-**21**
fill (of a cooling tower) *packing (of a cooling tower)*	**relleno (de una torre de enfriamiento)**	garnissage (d'un refroidisseur atmosphé-rique, d'une tour de refroidissement *remplissage (d'un refroidisseur atmosphé-rique, d'une tour de refroidissement)* *matelas dispersant*	2.5.2-**22**
film cooling tower	**torre de enfriamiento por corriente en película**	tour de refroidissement à écoulement pelliculaire	2.5.2-**23**
film packing	**relleno para corriente en película**	garnissage à écoulement pelliculaire	2.5.2-**24**
forced-draught condenser	**condensador de aire por convección forzada**	condenseur à air à convection forcée	2.5.2-**25**
forced-draught cooling tower	**torre de enfriamiento de agua con tiro forzado por impulsión** torre de enfriamiento de impulsión	refroidisseur d'eau atmosphérique à air forcé *tour de refroidissement à air forcé*	2.5.2-**26**
grid fill o *grid packing*	**relleno en emparrillado**	garnissage en treillage *garnissage en grille*	2.5.2-**27**
grid packing *grid fill* o	**relleno en emparrillado**	garnissage en treillage *garnissage en grille*	2.5.2-**28**
heat rejection capacity	**capacidad de eliminación de calor**	puissance thermique évacuée	2.5.2-**29**
hot water *warm water*	**agua caliente**	eau chaude	2.5.2-**30**
hot water basin *warm water basin*	**depósito de agua caliente**	bassin d'accumulation d'eau chaude	2.5.2-**31**
induced draught cooling tower	**torre de enfriamiento de agua con ventilación por aspiración** torre de enfriamiento de inducción torre de enfriamiento de aspiración	refroidisseur d'eau atmosphérique à ventilation par aspiration *tour de refroidissement à ventilation par aspiration* *refroidisseur d'eau atmosphérique à air induit* o *tour de refroidissement à air induit* o	2.5.2-**32**
liquefier	**licuefactor**	liquéfacteur	2.5.2-**33**

	ENGLISH	ESPAÑOL	FRANÇAIS
2.5.2-**34**	liquid-cooled refrigerant condenser	**condensador autónomo enfriado por líquido**	condenseur refroidi par un liquide
2.5.2-**35**	make-up water	**aportación de agua**	eau d'appoint
2.5.2-**36**	mechanical draught cooling tower	**torre de enfriamiento de circulación forzada** torre de enfriamiento de agua con tiro forzado	refroidisseur d'eau atmosphérique à ventilation forcée *tour de refroidissement à ventilation forcée*
2.5.2-**37**	multishell condenser ◐	**condensador multitubular de envolvente múltiple**	condenseur à calandres multiples *condenseur multitubulaire à plusieurs corps* ◐
2.5.2-**38**	(multitubular) shell-and-tube condenser ◐ *shell-and-tube condenser*	**condensador multitubular de envolvente**	condenseur à tubes et à calandre
2.5.2-**39**	natural-convection air-cooled condenser	**condensador de aire por convección natural**	condenseur à air statique
2.5.2-**40**	natural-draught cooling tower *atmospheric cooling tower* ○	**torre de enfriamiento de circulación natural** torre de enfriamiento de agua con tiro natural	refroidisseur d'eau atmosphérique à tirage naturel *tour de refroidissement à tirage naturel*
2.5.2-**41**	open shell-and-tube condenser	**condensador multitubular de envolvente abierto**	condenseur à tubes et calandre ouvert
2.5.2-**42**	packing (of a cooling tower) *fill (of a cooling tower)*	**relleno (de una torre de enfriamiento)**	garnissage (d'un refroidisseur atmosphérique, d'une tour de refroidissement) *remplissage (d'un refroidisseur atmosphérique, d'une tour de refroidissement)* *matelas dispersant*
2.5.2-**43**	plate condenser	**condensador de placas**	condenseur à plaques
2.5.2-**44**	plate fill ◐ *plate packing* ◐	**relleno con placas**	garnissage à plaques
2.5.2-**45**	plate packing ◐ *plate fill* ◐	**relleno con placas**	garnissage à plaques
2.5.2-**46**	random fill ◐ *random packing*	**relleno a granel**	garnissage en vrac *remplissage en vrac* ◐
2.5.2-**47**	random packing *random fill* ◐	**relleno a granel**	garnissage en vrac *remplissage en vrac* ◐
2.5.2-**48**	Raschig rings	**anillos Raschig**	anneaux Raschig
2.5.2-**49**	remote mechanical-draft air-cooled refrigerant condenser	**condensador autónomo por circulación forzada de aire**	condenseur à air à ventilation forcée autonome
2.5.2-**50**	scale inhibitor	**antiincrustante**	anti-tartre *anti-incrustant* ◐
2.5.2-**51**	shell-and-coil condenser	**condensador de serpentín de envolvente**	condenseur à calandre et serpentin *condenseur à virole et serpentin* ◐
2.5.2-**52**	shell-and-tube condenser *(multitubular) shell-and-tube condenser* ◐	**condensador multitubular de envolvente**	condenseur à tubes et à calandre
2.5.2-**53**	splash packing ◐	**relleno para pulverización**	garnissage dispersant
2.5.2-**54**	split condenser	**condensador de circuitos múltiples**	ensemble multicondenseurs
2.5.2-**55**	spray cooling tower	**torre de enfriamiento por pulverización**	tour de refroidissement à pulvérisation *tour de refroidissement sans garnissage* *refroidisseur d'eau à pulvérisation*
2.5.2-**56**	spray pond	**recipiente de enfriamiento de agua por aspersión**	refroidisseur d'eau à pulvérisation
2.5.2-**57**	subcooler	**subenfriador** (subst.)	sous-refroidisseur
2.5.2-**58**	submerged coil condenser ◐ *submerged condenser*	**condensador sumergido**	condenseur à immersion

ENGLISH	ESPAÑOL	FRANÇAIS	
submerged condenser *submerged coil condenser* ◐	**condensador sumergido**	condenseur à immersion	2.5.2-**59**
tube-in-tube condenser ◐	**condensador de doble tubo** condensador a contracorriente	condenseur à double tube	2.5.2-**60**
warm water *hot water*	**agua caliente**	eau chaude	2.5.2-**61**
warm water basin *hot water basin*	**depósito de agua caliente**	bassin d'accumulation d'eau chaude	2.5.2-**62**
water-cooled condenser	**condensador enfriado por agua**	condenseur à eau	2.5.2-**63**
welded plate condenser ◐	**condensador de placas soldadas**	condenseur à plaques soudées	2.5.2-**64**
wire-and-tube condenser	**condensador de tubos y alambres**	condenseur à tubes et fils	2.5.2-**65**

SECTION 2.5.3 *Evaporators and evaporator accessories*	**SUBCAPÍTULO 2.5.3** *Evaporadores y sus accesorios*	**SOUS-CHAPITRE 2.5.3** *Evaporateurs et accessoires*	
direct expansion evaporator *dry expansion evaporator*	**evaporador en régimen seco** evaporador de inyección directa	évaporateur à détente sèche	2.5.3-**1**
drier coil	**tubo secador**	tube sécheur *échangeur surchauffeur* ◐	2.5.3-**2**
dry expansion evaporator *direct expansion evaporator*	**evaporador en régimen seco** evaporador de inyección directa	évaporateur à détente sèche	2.5.3-**3**
embossed-plate evaporator ○	**evaporador de circuito estampado**	évaporateur à circuit embouti	2.5.3-**4**
evaporating unit	**unidad enfriadora**	groupe évaporateur *groupe d'évaporation* ◐	2.5.3-**5**
evaporator	**evaporador**	évaporateur	2.5.3-**6**
flooded evaporator	**evaporador inundado**	évaporateur noyé	2.5.3-**7**
forced-air-circulation unit air cooler ◐	**instalación frigorífica por convección forzada de aire**	aérofrigorifère ventilé *aérofrigorifère refroidisseur à convection forcée* ◐	2.5.3-**8**
herringbone-type evaporator ○ *v-coil evaporator*	**evaporador de tubos en V** evaporador de espina de pescado	évaporateur en chevrons *évaporateur en arêtes de poisson* ◐	2.5.3-**9**
ice bank evaporator	**evaporador acumulador de hielo**	évaporateur accumulateur de glace	2.5.3-**10**
integral plate evaporator	**evaporador de placas integrales**	évaporateur à plaques intégrales	2.5.3-**11**
liquid overfeed evaporator *recirculation-type evaporator*	**evaporador de recirculación**	évaporateur à recirculation	2.5.3-**12**
liquid suction heat interchanger *superheater* ◐	**cambiador de calor (entre fases de un fluido frigorífico)**	échangeur liquide-vapeur	2.5.3-**13**
nominal air flow	**flujo nominal de aire**	débit d'air nominal	2.5.3-**14**
plate coil *tube-on-sheet evaporator* ◐	**evaporador de placa con tubos exteriores**	évaporateur plaque et tubes	2.5.3-**15**
plate evaporator	**evaporador de placa con tubos interiores**	évaporateur platulaire	2.5.3-**16**
pump-fed evaporator	**evaporador alimentado por bomba**	évaporateur noyé alimenté par pompe	2.5.3-**17**
raceway coil ○	**evaporador de rastrillo**	évaporateur herse	2.5.3-**18**
recirculation-type evaporator *liquid overfeed evaporator*	**evaporador de recirculación**	évaporateur à recirculation	2.5.3-**19**
(refrigerant) distributor	**distribuidor (de fluido frigorífico)**	distributeur *distributeur de frigorigène* ◐	2.5.3-**20**

	ENGLISH	ESPAÑOL	FRANÇAIS
2.5.3-**21**	refrigerant recirculation rate	**caudal de recirculación del fluido frigorígeno**	taux de circulation
2.5.3-**22**	roll-bond evaporator	**evaporador soldado por laminación** evaporador "roll bond"	évaporateur "roll bond" *évaporateur soudé par laminage* o
2.5.3-**23**	shell-and-coil evaporator	**evaporador de serpentín y envolvente**	évaporateur à calandre et serpentin
2.5.3-**24**	shell-and-tube evaporator	**evaporador multitubular de envolvente**	évaporateur multitubulaire à calandre *évaporateur à tubes et calandre*
2.5.3-**25**	spray-type evaporator	**evaporador de aspersión**	évaporateur à aspersion interne
2.5.3-**26**	superheater o *liquid suction heat interchanger*	**cambiador de calor (entre fases de un fluido frigorífico)**	échangeur liquide-vapeur
2.5.3-**27**	tube-on-sheet evaporator o *plate coil*	**evaporador de placa con tubos exteriores**	évaporateur plaque et tubes
2.5.3-**28**	v-coil evaporator *herringbone-type evaporator* o	**evaporador de tubos en V** evaporador de espina de pescado	évaporateur en chevrons *évaporateur en arêtes de poisson* o
2.5.3-**29**	vertical shell-and-tube evaporator	**evaporador multitubular vertical**	évaporateur multitubulaire vertical
2.5.3-**30**	vertical-type evaporator	**evaporador de tubos verticales**	évaporateur à tubes verticaux
2.5.3-**31**	wrap-round evaporator	**evaporador envolvente**	évaporateur à double enveloppe

	SECTION 2.5.4 *Cooling distribution equipment*	SUBCAPÍTULO 2.5.4 *Equipos de distribución de frío*	SOUS-CHAPITRE 2.5.4 *Equipements de distribution de froid*
2.5.4-**1**	ceiling coil *overhead coil* o *ceiling grid* o	**serpentín de techo**	échangeur plafonnier
2.5.4-**2**	ceiling grid o *ceiling coil* *overhead coil* o	**serpentín de techo**	échangeur plafonnier
2.5.4-**3**	coil *grid* o	**serpentín**	serpentin *batterie* o
2.5.4-**4**	convector (equipment)	**convector (aparato)**	convecteur
2.5.4-**5**	cooling battery o	**batería fria** batería frigorífica	refroidisseur d'air *batterie frigorifique* o
2.5.4-**6**	cooling coil	**serpentín de enfriamiento**	serpentin refroidisseur *élément refroidisseur (tubulaire)* o
2.5.4-**7**	eutectic plate	**placa eutéctica**	plaque eutectique
2.5.4-**8**	grid o *coil*	**serpentín**	serpentin *batterie* o
2.5.4-**9**	hold-over coil	**acumulador de frío** serpentín acumulador (de frío)	serpentin accumulateur (de glace)
2.5.4-**10**	hold-over plate	**placa acumuladora de frío**	plaque accumulatrice (de froid)
2.5.4-**11**	overhead coil o *ceiling coil* *ceiling grid* o	**serpentín de techo**	échangeur plafonnier
2.5.4-**12**	wall coil *wall grid* o	**serpentín de pared**	échangeur mural
2.5.4-**13**	wall grid o *wall coil*	**serpentín de pared**	échangeur mural

ENGLISH	ESPAÑOL	FRANÇAIS	
SECTION 2.6 *Valves and pipes* **SECTION 2.6.1** *Valves: general background*	**SUBCAPÍTULO 2.6** *Valvulería y tuberías* **SUBCAPÍTULO 2.6.1** *Generalidades sobre valvulería*	**SOUS-CHAPITRE 2.6** *Robinetterie et tuyauteries* **SOUS-CHAPITRE 2.6.1** *Généralités sur la robinetterie*	
angle pattern body	**cuerpo (de válvula) en escuadra**	corps de robinet d'équerre	2.6.1-**1**
angle valve	**llave angular**	robinet d'équerre	2.6.1-**2**
ball cock ○ *ball (float) valve* ○	**válvula de flotador**	robinet à flotteur	2.6.1-**3**
ball (float) valve ○ *ball cock* ○	**válvula de flotador**	robinet à flotteur	2.6.1-**4**
ball (plug) valve ○ *plug-and-ball valve*	**válvula de bola** válvula esférica	robinet à tournant sphérique *robinet à boule* ○	2.6.1-**5**
ball valve	**válvula de bola** válvula esférica	clapet à bille	2.6.1-**6**
bellows seal	**cierre de fuelle**	soufflet d'étanchéité	2.6.1-**7**
bellows valve	**válvula de fuelle**	robinet à soufflet	2.6.1-**8**
body ○ *valve body*	**cuerpo de válvula** cuerpo (de la válvula)	corps de robinet *corps* ○	2.6.1-**9**
body end	**extremo del cuerpo**	extrémité du corps	2.6.1-**10**
body end port	**puerta de conexión del cuerpo** **(de la válvula)**	orifice d'extrémité de corps	2.6.1-**11**
body seat	**asiento del cuerpo (de la válvula)**	siège du corps	2.6.1-**12**
bonnet	**tapa**	tête de robinet	2.6.1-**13**
butterfly valve	**válvula de mariposa**	robinet à papillon	2.6.1-**14**
bypass valve *shunt valve* ○	**válvula de derivación**	robinet de dérivation *robinet de bipasse*	2.6.1-**15**
changeover device	**dispositivo de conmutación**	dispositif de commutation *inverseur-commutateur*	2.6.1-**16**
charging valve	**llave de carga**	robinet de charge	2.6.1-**17**
check valve *non-return valve* ○	**válvula de retención**	clapet de non-retour *clapet de retenue*	2.6.1-**18**
cock ○	**llave**	robinet (à liquide)	2.6.1-**19**
control valve	**válvula de regulación**	robinet de régulation *robinet régulateur (piloté)*	2.6.1-**20**
cover	**tapa**	couvercle	2.6.1-**21**
cover flange	**brida de ensamblaje (cuerpo-tapa)**	bride d'assemblage corps-chapeau ou corps-couvercle	2.6.1-**22**
cylinder valve	**llave para botella de fluido frigorígeno**	robinet de bouteille	2.6.1-**23**
diaphragm valve *membrane valve*	**válvula de membrana**	robinet à membrane	2.6.1-**24**
diverting valve	**válvula diversora**	robinet diviseur	2.6.1-**25**
double-flanged body	**cuerpo con bridas** llave con bridas	corps à brides	2.6.1-**26**
double-seated valve *two-way valve*	**válvula de doble asiento**	robinet de service à double siège	2.6.1-**27**
drain hole	**orificio de vaciado** orificio de purga	purgeur	2.6.1-**28**

	ENGLISH	ESPAÑOL	FRANÇAIS
2.6.1-**29**	drain valve	**llave de vaciado** llave de purga	robinet de soutirage *robinet de vidange*
2.6.1-**30**	effective port area	**sección efectiva de abertura**	section effective de passage
2.6.1-**31**	four-way valve	**válvula de cuatro vías**	robinet à quatre voies
2.6.1-**32**	gate valve	**llave de compuerta**	robinet vanne *vanne* ● *robinet à vanne* ○
2.6.1-**33**	globe valve	**válvula de globo**	robinet à soupape *robinet à clapet*
2.6.1-**34**	hand-stop valve	**llave manual de paso** válvula manual de cierre	robinet d'arrêt à main
2.6.1-**35**	initial valve opening	**abertura inicial de la válvula**	position minimale d'ouverture
2.6.1-**36**	isolating valve	**válvula de seccionamiento**	robinet de sectionnement
2.6.1-**37**	king valve (USA) *master valve*	**llave general** llave principal	robinet général
2.6.1-**38**	lift check valve	**válvula de retención de movimiento vertical**	clapet de retenue à mouvement vertical *clapet de retenue guidé*
2.6.1-**39**	lug-type body	**cuerpo con bridas**	corps à oreilles
2.6.1-**40**	mass flow controller	**válvula de retención** llave de retención	clapet d'arrêt
2.6.1-**41**	master valve *king valve (USA)*	**llave general** llave principal	robinet général
2.6.1-**42**	membrane valve *diaphragm valve*	**válvula de membrana**	robinet à membrane
2.6.1-**43**	mixing valve	**válvula mezcladora**	robinet mélangeur *robinet mitigeur* ●
2.6.1-**44**	multi-end body	**válvula con varias puertas**	corps multivoies
2.6.1-**45**	multiway plug-and-ball valve	**válvula esférica de salida múltiple**	robinet à tournant multivoies
2.6.1-**46**	needle valve	**válvula de aguja**	robinet à pointeau *robinet à aiguille*
2.6.1-**47**	non-return valve ● *check valve*	**válvula de retención**	clapet de non-retour *clapet de retenue*
2.6.1-**48**	oblique pattern body	**cuerpo inclinado**	corps à tête inclinée
2.6.1-**49**	obturator	**obturador**	obturateur
2.6.1-**50**	oil-charge valve	**llave de carga de aceite**	robinet de remplissage d'huile
2.6.1-**51**	oil-drain valve *oil-purge valve* ●	**llave de purga de aceite**	robinet de purge d'huile *robinet d'extraction d'huile*
2.6.1-**52**	oil-purge valve ● *oil-drain valve*	**llave de purga de aceite**	robinet de purge d'huile *robinet d'extraction d'huile*
2.6.1-**53**	operating mechanism	**mecanismo de accionamiento**	mécanisme de manoeuvre
2.6.1-**54**	packless valve	**llave sin empaquetadura** válvula sin empaquetadura	robinet sans garniture
2.6.1-**55**	permanent bleed-type valve	**válvula de purga permanente**	robinet à fuite permanente
2.6.1-**56**	piston valve *slide valve* ●	**válvula de pistón**	robinet à piston
2.6.1-**57**	plug-and-ball valve *ball (plug) valve* ●	**válvula de bola** válvula esférica	robinet à tournant sphérique *robinet à boule* ●

ENGLISH	ESPAÑOL	FRANÇAIS	
plug valve	**válvula obturadora**	robinet à tournant *robinet à boisseau* ○	2.6.1-**58**
port *valve area*	**sección de paso (de la válvula)**	section de passage (d'un robinet)	2.6.1-**59**
pump-out valve	**válvula de vacío** válvula de puesta en vacío	robinet de mise sous vide *robinet de mise à vide* ◐	2.6.1-**60**
purge valve	**llave de purga**	robinet de purge *robinet d'extraction* ◐	2.6.1-**61**
refrigerant access valve *Schrader valve*	**válvula de entrada del fluido frigorígeno** válvula de servicio y control	valve de service et de contrôle *valve Schrader*	2.6.1-**62**
refrigerant access valve hose connector	**conector flexible de acceso a la válvula**	flexible de raccordement à la valve Schrader	2.6.1-**63**
refrigerant valve core	**obús de la válvula**	noyau d'une valve de service ◐	2.6.1-**64**
regulating valve	**válvula de regulación**	robinet de réglage *robinet régleur* *régleur*	2.6.1-**65**
reversing valve	**válvula inversora**	robinet inverseur	2.6.1-**66**
Schrader valve *refrigerant access valve*	**válvula de entrada del fluido frigorígeno** válvula de servicio y control	valve de service et de contrôle *valve Schrader*	2.6.1-**67**
seating surface	**superficie de asiento**	portée d'étanchéité	2.6.1-**68**
service valve	**llave de servicio** válvula de servicio	robinet de service	2.6.1-**69**
shell	**envolvente**	corps	2.6.1-**70**
shell tapping	**orificio roscado en la envolvente**	raccordement auxiliaire sur le corps	2.6.1-**71**
shunt valve ◐ *bypass valve*	**válvula de derivación**	robinet de dérivation *robinet de bipasse*	2.6.1-**72**
shut-off valve ◐ *stop valve* ◐	**llave de paso** llave de cierre válvula de cierre	robinet d'arrêt	2.6.1-**73**
single-flanged body	**llave de brida única**	corps monobride	2.6.1-**74**
slide valve ◐ *piston valve*	**válvula de pistón**	robinet à piston	2.6.1-**75**
stop valve ◐ *shut-off valve* ◐	**llave de paso** llave de cierre válvula de cierre	robinet d'arrêt	2.6.1-**76**
straight-pattern body	**cuerpo recto**	corps à tête droite	2.6.1-**77**
swing-check valve	**válvula de retención de batiente**	clapet de retenue à battant	2.6.1-**78**
tap	**llave de extracción**	robinet de puisage	2.6.1-**79**
taper seat	**asiento oblicuo**	siège oblique	2.6.1-**80**
three-way valve	**válvula de tres vías**	robinet à trois voies	2.6.1-**81**
throttle valve	**válvula de estrangulación**	robinet d'étranglement	2.6.1-**82**
two-way valve *double-seated valve*	**válvula de doble asiento**	robinet de service à double siège	2.6.1-**83**
valve area *port*	**sección de paso (de la válvula)**	section de passage (d'un robinet)	2.6.1-**84**
valve body *body* ◐	**cuerpo de válvula** cuerpo (de la válvula)	corps de robinet *corps* ◐	2.6.1-**85**
valve cap	**tapa roscada estanca**	chapeau	2.6.1-**86**
valve plug	**obturador de la válvula**	pointeau	2.6.1-**87**

ENGLISH	ESPAÑOL	FRANÇAIS
2.6.1-88 valve trim	**mecanismo de la válvula**	mécanisme du robinet
2.6.1-89 valve (in general)	**válvula** llave	robinet (en général)
2.6.1-90 vent hole	**respiradero**	évent

SECTION 2.6.2 *Automatic valves*	SUBCAPÍTULO 2.6.2 *Válvulas automáticas*	SOUS-CHAPITRE 2.6.2 *Robinets automatiques*
2.6.2-1 automatic steam trap	**purgador automático de condensado de vapor**	purgeur automatique de vapeur d'eau
2.6.2-2 automatic valve	**válvula automática**	robinet automatique *détendeur automatique* ◑
2.6.2-3 back-pressure regulator *evaporator pressure regulator* *back-pressure valve* *constant-pressure valve*	**válvula de control de la presión de evaporación**	vanne de régulation de la pression d'évaporation
2.6.2-4 back-pressure valve *evaporator pressure regulator* *back-pressure regulator* *constant-pressure valve*	**válvula de control de la presión de evaporación**	vanne de régulation de la pression d'évaporation
2.6.2-5 condensing-pressure valve *high-pressure valve*	**válvula de control de la presión de condensación**	vanne de régulation de la pression de condensation
2.6.2-6 constant-pressure valve *evaporator pressure regulator* *back-pressure regulator* *back-pressure valve*	**válvula de control de la presión de evaporación**	vanne de régulation de la pression d'évaporation
2.6.2-7 evaporator pressure regulator *back-pressure regulator* *back-pressure valve* *constant-pressure valve*	**válvula de control de la presión de evaporación**	vanne de régulation de la pression d'évaporation
2.6.2-8 high-pressure valve *condensing-pressure valve*	**válvula de control de la presión de condensación**	vanne de régulation de la pression de condensation
2.6.2-9 hold-back valve ◑ *suction-pressure regulator*	**válvula de control de la presión de aspiración**	vanne de régulation de la pression d'aspiration
2.6.2-10 hot gas bypass regulator	**válvula de derivación de gases calientes**	(robinet) régulateur de dérivation des gaz chauds *robinet de bipasse des gaz chauds*
2.6.2-11 hydraulically actuated valve	**válvula hidráulica**	robinet à commande hydraulique
2.6.2-12 liquid-injection valve	**válvula de inyección de líquido**	robinet d'injection de liquide *détendeur d'injection*
2.6.2-13 magnetic valve *solenoid valve*	**válvula de solenoide** válvula electromagnética	robinet électromagnétique *robinet solénoïde* *électrovanne* ◑ *robinet d'arrêt électromagnétique* ◑
2.6.2-14 modulating valve	**válvula moduladora**	robinet modulant
2.6.2-15 motor-operated valve *motorized valve* ◑	**válvula motorizada**	robinet motorisé

ENGLISH	ESPAÑOL	FRANÇAIS	
motorized valve ◐ *motor-operated valve*	**válvula motorizada**	robinet motorisé	2.6.2-**16**
pilot-controlled valve *servo-operated valve ◐*	**válvula servoaccionada** válvula servocomandada	robinet à servocommande	2.6.2-**17**
pilot valve	**válvula piloto**	robinet pilote	2.6.2-**18**
pneumatic-operated valve	**válvula neumática**	robinet pneumatique	2.6.2-**19**
pressure control valve	**válvula reguladora de presión**	vanne de régulation de pression	2.6.2-**20**
pressure-controlled valve	**válvula presostática**	robinet pressostatique	2.6.2-**21**
(pressure) reducing valve	**válvula reductora de presión**	robinet réducteur de pression	2.6.2-**22**
quick-closing valve	**válvula de cierre rápido**	robinet à fermeture rapide *soupape à fermeture rapide*	2.6.2-**23**
servo-operated valve ◐ *pilot-controlled valve*	**válvula servoaccionada** válvula servocomandada	robinet à servocommande	2.6.2-**24**
snap-action valve	**barostato** regulador de presión de acción instantánea	régulateur de pression à déclic	2.6.2-**25**
solenoid valve *magnetic valve*	**válvula de solenoide** válvula electromagnética	robinet électromagnétique *robinet solénoïde* *électrovanne ◐* *robinet d'arrêt électromagnétique ◐*	2.6.2-**26**
suction-pressure regulator *hold-back valve ◐*	**válvula de control de la presión de aspiración**	vanne de régulation de la pression d'aspiration	2.6.2-**27**
thermostatically controlled valve	**válvula termostática**	robinet thermostatique	2.6.2-**28**
water-regulating valve	**válvula de regulación del caudal de agua**	(robinet) régulateur de débit d'eau	2.6.2-**29**

SECTION 2.6.3 *Expansion devices*	SUBCAPÍTULO 2.6.3 *Dispositivos de expansión*	SOUS-CHAPITRE 2.6.3 *Détendeurs*	
adjustable capillary valve ○	**válvula capilar regulable**	robinet capillaire réglable	2.6.3-**1**
capacity of an expansion valve	**potencia frigorífica del dispositivo de expansión**	capacité frigorifique d'un détendeur *puissance frigorifique d'un détendeur ◐*	2.6.3-**2**
capillary tube	**tubo capilar** capilar (subst.)	capillaire	2.6.3-**3**
compensator	**compensador** (subst.)	dispositif d'affranchissement de l'ambiance	2.6.3-**4**
direct-acting valve	**válvula de acción directa**	détendeur thermostatique à action directe	2.6.3-**5**
expansion device	**dispositivo de expansión**	dispositif de détente	2.6.3-**6**
expansion valve	**válvula de expansión** válvula de laminación válvula de regulación	détendeur *robinet détendeur ◐*	2.6.3-**7**
external equalizer	**equilibrador exterior**	égalisateur externe	2.6.3-**8**
float-type expansion valve ○ *float valve ◐*	**válvula (de expansión) de flotador**	régleur à flotteur ◐ *détendeur à flotteur ○*	2.6.3-**9**
float valve ◐ *float-type expansion valve ○*	**válvula (de expansión) de flotador**	régleur à flotteur ◐ *détendeur à flotteur ○*	2.6.3-**10**
gas-charged thermostat *limited liquid charged thermostat*	**termostato de carga líquida limitada** termostato de carga de vapor	thermostat à charge (liquide) limitée *thermostat chargé en vapeur*	2.6.3-**11**
hand expansion valve	**válvula de expansión manual**	détendeur à main	2.6.3-**12**

	ENGLISH	ESPAÑOL	FRANÇAIS
2.6.3-**13**	high-pressure float valve *high-side float valve (USA)* ⊙	**válvula de flotador de alta presión**	détendeur à flotteur haute pression *régleur à flotteur haute pression* ⊙
2.6.3-**14**	high-side float valve (USA) ⊙ *high-pressure float valve*	**válvula de flotador de alta presión**	détendeur à flotteur haute pression *régleur à flotteur haute pression* ⊙
2.6.3-**15**	internal equalizer	**equilibrador interno**	égalisateur interne
2.6.3-**16**	limited liquid charged thermostat *gas-charged thermostat*	**termostato de carga líquida limitada** termostato de carga de vapor	thermostat à charge (liquide) limitée *thermostat chargé en vapeur*
2.6.3-**17**	liquid-charged thermostat	**termostato con carga líquida**	thermostat chargé en liquide *thermostat chargé en vapeur saturée-liquide* *thermostat à charge liquide*
2.6.3-**18**	low-pressure float valve *low-side float valve (USA)* ⊙	**válvula de flotador de baja presión**	régleur à flotteur basse pression
2.6.3-**19**	low-side float valve (USA) ⊙ *low-pressure float valve*	**válvula de flotador de baja presión**	régleur à flotteur basse pression
2.6.3-**20**	maximum operating pressure (MOP)	**presión máxima de funcionamiento (PMF)**	pression maximale de fonctionnement
2.6.3-**21**	multiport expansion valve	**válvula de expansión de orificios múltiples** válvula de expansión de salida múltiple	détendeur à orifices multiples *détendeur multi-orifices*
2.6.3-**22**	operating superheat	**sobrecalentamiento de funcionamiento** sobrecalentamiento	surchauffe de fonctionnement *surchauffe opérationnelle*
2.6.3-**23**	pilot-operated expansion valve	**válvula piloto**	(robinet) détendeur piloté
2.6.3-**24**	power element (of a thermostat) *power system (of a thermostat)* ⊙	**tren termostático**	train thermostatique *actionneur*
2.6.3-**25**	power system (of a thermostat) ⊙ *power element (of a thermostat)*	**tren termostático**	train thermostatique *actionneur*
2.6.3-**26**	refrigerant metering device	**dispositivo de regulación (de caudal de fluido)** regulador	appareil de réglage de débit de fluide *régleur*
2.6.3-**27**	restrictor	**restrictor** (subst.)	restricteur
2.6.3-**28**	restrictor valve	**válvula de restricción**	dispositif restricto-changeur
2.6.3-**29**	static superheat	**umbral de sobrecalentamiento**	seuil de surchauffe
2.6.3-**30**	superheat change	**sobrecalentamiento efectivo**	surchauffe effective
2.6.3-**31**	temperature-sensing element	**bulbo del tren termostático**	bulbe du train thermostatique
2.6.3-**32**	(thermostatic) adsorber charge	**carga adsorbente (termostática)**	charge adsorbante (thermostatique)
2.6.3-**33**	(thermostatic) cross charge	**carga heterogénea (termostática)** carga cruzada	hétérocharge (thermostatique)
2.6.3-**34**	(thermostatic) dry charge ⊙ *(thermostatic) gas charge* *(thermostatic) limited liquid charge* ⊙	**carga limitada (termostática)**	charge limitée (thermostatique)
2.6.3-**35**	thermostatic expansion valve	**válvula de expansión termostática**	détendeur thermostatique
2.6.3-**36**	thermostatic expansion valve charge	**carga de la válvula de expansión termostática**	charge de détendeur thermostatique
2.6.3-**37**	thermostatic expansion valve pressure drop	**caída de presión en la válvula de expansión termostática**	chute de pression à travers le détendeur
2.6.3-**38**	thermostatic expansion valve superheat	**sobrecalentamiento de la válvula de expansión termostática**	surchauffe d'un détendeur thermo-statique
2.6.3-**39**	(thermostatic) gas charge *(thermostatic) dry charge* ⊙ *(thermostatic) limited liquid charge* ⊙	**carga limitada (termostática)**	charge limitée (thermostatique)

ENGLISH	ESPAÑOL	FRANÇAIS	
(thermostatic) limited liquid charge ⊙ *(thermostatic) gas charge* *(thermostatic) dry charge* ⊙	**carga limitada (termostática)**	charge limitée (thermostatique)	2.6.3-**40**
(thermostatic) liquid charge *(thermostatic) wet charge* ⊙	**carga líquida (termostática)**	charge liquide (thermostatique)	2.6.3-**41**
(thermostatic) straight charge ⊙	**carga homogénea (termostática)**	homocharge (thermostatique)	2.6.3-**42**
(thermostatic) wet charge ⊙ *(thermostatic) liquid charge*	**carga líquida (termostática)**	charge liquide (thermostatique)	2.6.3-**43**

SECTION 2.6.4 *Piping and fittings*	SUBCAPÍTULO 2.6.4 *Tuberías y accesorios*	SOUS-CHAPITRE 2.6.4 *Tuyauteries et raccords*	
braze-welding	**soldadura (con metal de aporte) progressiva**	soudobrasage	2.6.4-**1**
brazing	**soldadura (con metal de aporte) progressiva**	brasage	2.6.4-**2**
brine line	**conducto de salmuera** línea de salmuera	conduite de saumure	2.6.4-**3**
butt welding end	**soldadura a tope**	extrémité à souder en bout	2.6.4-**4**
bypass	**derivación** bypass	bipasse *dérivation* ⊙	2.6.4-**5**
cap	**tapón hembra**	bouchon femelle *capuchon*	2.6.4-**6**
capillary end	**extremo para soldadura por capilaridad**	extrémité à braser par capillarité	2.6.4-**7**
capillary fitting *sweat joint* ⊙	**junta soldada (por capilaridad)**	joint brasé par capillarité	2.6.4-**8**
charging connection	**conexión de carga** racor de carga	raccord de charge	2.6.4-**9**
clamp ring	**anillo de abrazadera**	collier de serrage	2.6.4-**10**
coil of tubing	**rollo de tubo**	couronne de tube	2.6.4-**11**
condensate line	**conducto del condensado**	conduite de condensat	2.6.4-**12**
connecting hose *hose assembly* ⊙	**unión flexible**	flexible (de raccordement)	2.6.4-**13**
coupling	**manguito de unión**	manchon (de raccordement)	2.6.4-**14**
discharge line *hot gas line* ⊙	**conducto de impulsión** línea de impulsión tubería de impulsión	conduite de refoulement *tuyauterie de refoulement*	2.6.4-**15**
double (suction) riser	**conducto ascendente doble**	conduite ascendante double	2.6.4-**16**
double male reduction	**reducción macho-macho**	réduction double mâle	2.6.4-**17**
expansion bend	**lira de dilatación**	lyre de dilatation	2.6.4-**18**
expansion joint	**junta de dilatación**	joint de dilatation	2.6.4-**19**
expansion loop ⊙	**codo de dilatación**	boucle de dilatation *cor de chasse de dilatation*	2.6.4-**20**
fittings	**accesorios**	raccords	2.6.4-**21**
flange	**brida**	bride	2.6.4-**22**
flanged end	**extremo embridado**	extrémité à bride	2.6.4-**23**

	ENGLISH	ESPAÑOL	FRANÇAIS
2.6.4-**24**	flanged joint	**junta embridada**	joint à bride
2.6.4-**25**	flare fitting *flared joint*	**junta mandrinada** junta cónica	collet *joint conique* *joint évasé* *"dudgeon"* o
2.6.4-**26**	flared joint *flare fitting*	**junta mandrinada** junta cónica	collet *joint conique* *joint évasé* *"dudgeon"* o
2.6.4-**27**	flaring block o *flaring tool*	**abocardador**	dudgeonnière *outil à façonner les collets*
2.6.4-**28**	flaring tool *flaring block* o	**abocardador**	dudgeonnière *outil à façonner les collets*
2.6.4-**29**	flexible pipe element	**elemento de conducto flexible**	élément flexible de tuyauterie
2.6.4-**30**	gasket	**empaquetadura** material para junta	joint (matériau) *garniture* o
2.6.4-**31**	hard soldering	**soldadura fuerte**	brasage fort
2.6.4-**32**	header *manifold*	**colector** colector múltiple	collecteur (à raccordements multiples) *manifold* o
2.6.4-**33**	heavy gauge	**de gran espesor**	de forte épaisseur
2.6.4-**34**	hose assembly o *connecting hose*	**unión flexible**	flexible (de raccordement)
2.6.4-**35**	hot gas defrost line	**tubería de desescarche por gases calientes**	conduite de gaz chaud pour dégivrage
2.6.4-**36**	hot gas line o *discharge line*	**conducto de impulsión** línea de impulsión tubería de impulsión	conduite de refoulement *tuyauterie de refoulement*
2.6.4-**37**	joint (in general)	**junta (en general)**	joint (en général)
2.6.4-**38**	joint ring ○	**junta tórica** junta anular	joint torique *joint annulaire* o
2.6.4-**39**	line *main* o	**conducto** tubería canalización	canalisation *conduite* *tuyauterie* o
2.6.4-**40**	liquid line	**linea de líquido** tubería de líquido	tuyauterie de liquide *conduite de liquide* o
2.6.4-**41**	main o *line*	**conducto** tubería canalización	canalisation *conduite* *tuyauterie* o
2.6.4-**42**	main pipe	**conducto principal**	canalisation principale *conduite principale*
2.6.4-**43**	male-female facing flange *raised-face flange* o *R-F flange* o	**brida de enchufe simple**	bride à emboîtement simple *bride à face surélevée*
2.6.4-**44**	manifold *header*	**colector** colector múltiple	collecteur (à raccordements multiples) *manifold* o
2.6.4-**45**	mechanical joint	**junta mecánica**	joint mécanique
2.6.4-**46**	metallic flexible pipe	**conducto metálico flexible**	tuyauterie flexible métallique
2.6.4-**47**	nipple	**niple**	mamelon
2.6.4-**48**	O-ring joint	**junta tórica de sección circular** junta anular de sección circular	joint torique (à section circulaire)
2.6.4-**49**	pipe schedule	**especificación de tubos**	spécification des tubes

ENGLISH	ESPAÑOL	FRANÇAIS	
piping	**tuberías**	tuyauterie	2.6.4-**50**
plug	**tapón macho**	bouchon mâle	2.6.4-**51**
quick-coupling *quick-release coupling* ●	**racor rápido**	raccord instantané *raccord rapide*	2.6.4-**52**
quick-release coupling ● *quick-coupling*	**racor rápido**	raccord instantané *raccord rapide*	2.6.4-**53**
R-F flange ● *male-female facing flange* *raised-face flange* ●	**brida de enchufe simple**	bride à emboîtement simple *bride à face surélevée*	2.6.4-**54**
raised-face flange ● *male-female facing flange* *R-F flange* ●	**brida de enchufe simple**	bride à emboîtement simple *bride à face surélevée*	2.6.4-**55**
reducing bushing	**reducción macho-hembra**	réduction mâle-femelle	2.6.4-**56**
reducing coupling	**reducción hembra-hembra**	réduction femelle-femelle *réduction double femelle*	2.6.4-**57**
return line ● *suction line*	**linea de aspiración** tubería de aspiración	tuyauterie d'aspiration *canalisation d'aspiration* ● *conduite d'aspiration* ●	2.6.4-**58**
screwed joint	**junta roscada**	joint vissé	2.6.4-**59**
sealant	**compuesto sellante**	produits d'étanchéité	2.6.4-**60**
seam	**junta longitudinal**	couture	2.6.4-**61**
seamless pipe	**tubo (estirado) sin soldadura**	tube (étiré) sans soudure	2.6.4-**62**
socket female end	**terminal de conexión**	extrémité à emboîter femelle *embout femelle*	2.6.4-**63**
socket welding end	**terminal (preparado) para soldar un componente**	extrémité à emboîter et à souder *embout à souder*	2.6.4-**64**
(soft) soldering	**soldadura blanda**	brasage tendre	2.6.4-**65**
soldered joint	**junta soldada**	joint brasé tendre	2.6.4-**66**
spool piece	**racor de bridas**	entretoise à brides *raccord à brides* ● *manchette* ●	2.6.4-**67**
suction line *return line* ●	**linea de aspiración** tubería de aspiración	tuyauterie d'aspiration *canalisation d'aspiration* ● *conduite d'aspiration* ●	2.6.4-**68**
swaging tool	**estampadora** embutidora	appareil à façonner les emboîtures *appareil à mandriner* ●	2.6.4-**69**
sweat joint ● *capillary fitting*	**junta soldada (por capilaridad)**	joint brasé par capillarité	2.6.4-**70**
taper pipe thread end	**rosca cónica de tubo**	joint fileté conique	2.6.4-**71**
threaded end	**terminal roscado**	extrémité filetée	2.6.4-**72**
tongue-and-groove facing flange ●	**brida de enchufe doble**	bride à emboîtement double	2.6.4-**73**
vibration isolator	**antivibrador**	isolateur de vibration	2.6.4-**74**
weld neck flange	**brida de collarín**	bride à collerette (à souder en bout)	2.6.4-**75**
welded joint	**junta soldada**	joint soudé	2.6.4-**76**
welded tube	**tubo soldado**	tube soudé	2.6.4-**77**
welding	**soldadura autógena**	soudage autogène	2.6.4-**78**
welding end	**terminal para soldar**	embout à souder	2.6.4-**79**

ENGLISH	ESPAÑOL	FRANÇAIS
SECTION 2.7 *Other equipment used for refrigeration production*	**SUBCAPÍTULO 2.7** *Otro material utilizado en instalaciones de producción de frío*	**SOUS-CHAPITRE 2.7** *Autres équipements utilisés pour la production de froid*
2.7-1 accumulator	**acumulador**	réservoir tampon *bouteille tampon* ◐
2.7-2 back-pressure gauge ○ *low-pressure gauge* *suction gauge*	**manómetro de aspiración** manómetro de baja (presión)	manomètre basse pression (B.P.) *manomètre d'aspiration*
2.7-3 balance tank	**recipiente de equilibrio**	réservoir d'équilibre
2.7-4 buffer tank	**recipiente regulador**	réservoir tampon
2.7-5 centrifugal pump	**bomba centrífuga**	pompe centrifuge
2.7-6 chilled water jacketed tank	**tanque de doble pared con circulación de agua fría**	bac à enveloppe d'eau glacée
2.7-7 circulation pump	**bomba de circulación**	pompe de circulation
2.7-8 compartment	**compartimento**	compartiment
2.7-9 component	**componente**	composant
2.7-10 dead end trap	**trampa en la aspiración**	piège à l'aspiration
2.7-11 dehydrator *drier*	**deshidratador** (subst.) secador (subst.)	déshydrateur *dessiccateur* ◐ *sécheur* ◐
2.7-12 desiccant	**substancia deshidratadora** desecante	déshydratant
2.7-13 discharge gauge (UK) discharge gage (USA) *high-pressure gauge* *head-pressure gauge* ◐	**manómetro de impulsión** manómetro de alta (presión)	manomètre de refoulement *manomètre haute pression*
2.7-14 drier *dehydrator*	**deshidratador** (subst.) secador (subst.)	déshydrateur *dessiccateur* ◐ *sécheur* ◐
2.7-15 dump trap liquid return	**retorno del líquido a alta presión**	retour de liquide à la haute pression
2.7-16 equalizer	**equilibrador (subst.)**	égalisateur (de pression)
2.7-17 equalizer tank	**recipiente de equilibrio**	réservoir égalisateur
2.7-18 filter *strainer* ◐ *screen* ○	**filtro**	filtre
2.7-19 filter-dehydrator ◐ *filter-drier*	**filtro deshidratador** deshidratador con filtro	déshydrateur-filtre
2.7-20 filter-drier *filter-dehydrator* ◐	**filtro deshidratador** deshidratador con filtro	déshydrateur-filtre
2.7-21 flash chamber	**cámara de separación (detrás de la válvula de expansión)**	chambre de séparation (après détendeur)
2.7-22 flash intercooler	**enfriador intermedio de expansión**	refroidisseur intermédiaire à détente
2.7-23 frost level indicator	**nivel de escarche**	niveau à givrage
2.7-24 gas purger ◐ *non-condensable gas purger*	**purgador de aire** purgador de gases no condensables	désaérateur *purgeur de gaz non condensables* *dégazeur* ◐
2.7-25 gear pump	**bomba de engranajes**	pompe à engrenage

ENGLISH	ESPAÑOL	FRANÇAIS	
head-pressure gauge ◐ *discharge gauge (UK) discharge gage (USA)* *high-pressure gauge*	**manómetro de impulsión** manómetro de alta (presión)	manomètre de refoulement *manomètre haute pression*	2.7-**26**
high-pressure gauge *discharge gauge (UK) discharge gage (USA)* *head-pressure gauge ◐*	**manómetro de impulsión** manómetro de alta (presión)	manomètre de refoulement *manomètre haute pression*	2.7-**27**
high-side receiver *receiver* *liquid receiver*	**recipiente de líquido**	réservoir de liquide haute pression *bouteille accumulatrice de liquide haute* *pression*	2.7-**28**
intercooler *interstage cooler ◐*	**enfriador intermedio**	refroidisseur intermédiaire	2.7-**29**
interstage cooler ◐ *intercooler*	**enfriador intermedio**	refroidisseur intermédiaire	2.7-**30**
liquid flow indicator	**visor de líquido**	voyant (de) liquide *contrôleur de circulation* *indicateur de passage de liquide*	2.7-**31**
liquid level indicator	**indicador de nivel (de líquido)**	indicateur de niveau (de liquide)	2.7-**32**
liquid pocket	**bolsa de líquido**	poche de liquide	2.7-**33**
liquid receiver *receiver* *high-side receiver*	**recipiente de líquido**	réservoir de liquide haute pression *bouteille accumulatrice de liquide haute* *pression*	2.7-**34**
liquid separator ◐ *suction accumulator ◐* *suction trap ◐*	**separador de líquido**	séparateur de liquide *bouteille anti-coup de liquide*	2.7-**35**
liquid trap	**dispositivo de retención de líquido**	piège à liquide	2.7-**36**
low-pressure gauge *suction gauge* *back-pressure gauge ○*	**manómetro de aspiración** manómetro de baja (presión)	manomètre basse pression (B.P.) *manomètre d'aspiration*	2.7-**37**
low-pressure receiver *surge drum ◐* *surge tank ◐*	**recipiente separador de líquido** separador de líquido de baja presión	réservoir de liquide basse pression *ballon basse pression*	2.7-**38**
non-condensable gas purger *gas purger ◐*	**purgador de aire** purgador de gases no condensables	désaérateur *purgeur de gaz non condensables* *dégazeur ◐*	2.7-**39**
oil drain	**purgador de aceite**	purgeur d'huile	2.7-**40**
oil pocket *oil slug*	**bolso de aceite** bolsa de aceite	poche d'huile	2.7-**41**
oil receiver	**recipiente de aceite**	récepteur d'huile	2.7-**42**
oil rectifier *oil still ◐*	**rectificador de aceite**	rectificateur d'huile	2.7-**43**
oil separator	**separador de aceite**	séparateur d'huile *déshuileur*	2.7-**44**
oil slug *oil pocket*	**bolso de aceite** bolsa de aceite	poche d'huile	2.7-**45**
oil still ◐ *oil rectifier*	**rectificador de aceite**	rectificateur d'huile	2.7-**46**
oil trap	**dispositivo de retención de aceite**	piège à huile	2.7-**47**
pressure equipment	**equipo de presión** presurizador	équipement sous pression	2.7-**48**
pressure vessel	**recipiente a presión**	récipient sous pression	2.7-**49**

ENGLISH	ESPAÑOL	FRANÇAIS
2.7-**50**	purge recovery system *purge unit* *purging device*	**grupo de purga** dispositivo de purga	groupe de purge *florentin* ●
2.7-**51**	purge unit *purge recovery system* *purging device*	**grupo de purga** dispositivo de purga	groupe de purge *florentin* ●
2.7-**52**	purging device *purge recovery system* *purge unit*	**grupo de purga** dispositivo de purga	groupe de purge *florentin* ●
2.7-**53**	receiver *high-side receiver* *liquid receiver*	**recipiente de líquido**	réservoir de liquide haute pression *bouteille accumulatrice de liquide haute pression*
2.7-**54**	scale trap	**separador de impurezas**	séparateur d'impuretés
2.7-**55**	screen ○ *filter* *strainer* ●	**filtro**	filtre
2.7-**56**	sight glass	**visor** mirilla	voyant *regard*
2.7-**57**	strainer ● *filter* *screen* ○	**filtro**	filtre
2.7-**58**	suction accumulator ● *liquid separator* ● *suction trap* ●	**separador de líquido**	séparateur de liquide *bouteille anti-coup de liquide*
2.7-**59**	suction gauge *low-pressure gauge* *back-pressure gauge* ○	**manómetro de aspiración** manómetro de baja (presión)	manomètre basse pression (B.P.) *manomètre d'aspiration*
2.7-**60**	suction trap ● *liquid separator* ● *suction accumulator* ●	**separador de líquido**	séparateur de liquide *bouteille anti-coup de liquide*
2.7-**61**	surge drum ● *low-pressure receiver* *surge tank* ●	**recipiente separador de líquido** separador de líquido de baja presión	réservoir de liquide basse pression *ballon basse pression*
2.7-**62**	surge tank ● *low-pressure receiver* *surge drum* ●	**recipiente separador de líquido** separador de líquido de baja presión	réservoir de liquide basse pression *ballon basse pression*

SECTION 2.8 *Control – Safety devices*	SUBCAPÍTULO 2.8 *Regulación – Control – Seguridad*	SOUS-CHAPITRE 2.8 *Régulation – Contrôle – Sécurité*
2.8-**1**	actuator	**actuador**	actionneur
2.8-**2**	adjustability	**banda de ajuste**	plage de réglage *possibilité de réglage* ●
2.8-**3**	adjustment	**reglaje** ajuste	réglage *mise au point* ●
2.8-**4**	automatic control	**accionamiento automático** mando automático	commande automatique
2.8-**5**	automatic control device	**aparato de regulación automática** dispositivo de control automático	appareil de régulation automatique
2.8-**6**	automatic control engineering	**automática** (subst.)	automatique (subst.)
2.8-**7**	automatic operation	**funcionamiento automático**	automatisme

ENGLISH	ESPAÑOL	FRANÇAIS	
automatic sequence control ◐	**funcionamiento automático cíclico**	automatisme séquentiel	2.8-**8**
automation	**automatización**	automatisation	2.8-**9**
bursting disc *rupture disc* *safety disc* ◐ *frangible disc* ○	**disco de seguridad** disco de ruptura	disque de rupture *disque de sûreté* *membrane d'éclatement*	2.8-**10**
cascade control system	**regulación por fases** regulación en cascada	régulation en cascade	2.8-**11**
change-over switch *selector switch* ◐	**commutador**	commutateur	2.8-**12**
closed loop *feedback loop*	**bucle de regulación**	boucle de régulation	2.8-**13**
control	**regulación (en general)** mando control	commande *régulation (en général)*	2.8-**14**
control console	**consola de mandos** pupitre de mandos	pupitre de commande *console de commande* ◐	2.8-**15**
control panel	**cuadro de mandos**	armoire de commande *coffret de commande* ◐ *tableau de commande* ◐	2.8-**16**
control rate	**velocidad de regulación**	vitesse de régulation *délai de réponse* *temps de réponse*	2.8-**17**
control thermostat	**termostato de control**	thermostat de régulation	2.8-**18**
controller	**dispositivo de accionamiento** dispositivo de regulación regulador	régulateur *dispositif de commande* ◐ *dispositif de régulation* ◐	2.8-**19**
counterweight safety valve	**válvula de seguridad de contrapeso**	soupape de sûreté à contrepoids	2.8-**20**
cut-in-point	**punto de conexión**	point d'enclenchement	2.8-**21**
cut-out point	**punto de desconexión**	point de coupure *point de déclenchement*	2.8-**22**
damping	**amortiguación**	amortissement	2.8-**23**
dead band *dead zone* ◐	**zona neutra**	zone d'insensibilité *plage d'insensibilité* *zone morte*	2.8-**24**
dead time *lag time*	**tiempo muerto**	temps mort	2.8-**25**
dead zone ◐ *dead band*	**zona neutra**	zone d'insensibilité *plage d'insensibilité* *zone morte*	2.8-**26**
detecting element *sensor* *sensing element* ◐	**sensor** detector elemento sensible	capteur *élément sensible*	2.8-**27**
differential controller	**regulador diferencial**	régulateur différentiel *appareil de régulation à différentiel* ◐	2.8-**28**
differential of a controller	**diferencial de un aparato regulador**	différentiel d'un régulateur *fourchette d'un régulateur*	2.8-**29**
dual pressure controller	**presostato combinado de alta y baja presión**	pressostat combiné haute pression-basse pression	2.8-**30**
feedback control	**regulación (en sentido estricto)** regulación por retroacción	commande par rétroaction	2.8-**31**

ENGLISH	ESPAÑOL	FRANÇAIS
2.8-**32**	feedback loop *closed loop*	**bucle de regulación**	boucle de régulation
2.8-**33**	feeler (bulb) ○ *thermostat bulb* *sensing bulb* ● *thermostat vial* ○	**bulbo termostático** bulbo sensible	bulbe (sensible) *bulbe (thermostatique)*
2.8-**34**	float switch	**interruptor de flotador**	interrupteur à flotteur
2.8-**35**	frangible disc ○ *bursting disc* *rupture disc* *safety disc* ●	**disco de seguridad** disco de ruptura	disque de rupture *disque de sûreté* *membrane d'éclatement*
2.8-**36**	fusible plug	**fusible** tapón fusible	fusible *bouchon fusible*
2.8-**37**	graduated acting	**modulante**	modulant *à action progressive*
2.8-**38**	halide torch	**lámpara haloidea** lámpara halógena	lampe haloïde
2.8-**39**	high discharge temperature cut-out	**termostato de seguridad de impulsión**	thermostat de sécurité de refoulement
2.8-**40**	high-low action ● *high-low control*	**regulación por todo o poco**	commande par tout ou peu
2.8-**41**	high-low control *high-low action* ●	**regulación por todo o poco**	commande par tout ou peu
2.8-**42**	high-pressure controller	**presostato de alta presión**	pressostat haute pression
2.8-**43**	high-pressure safety cut-out	**presostato de seguridad de alta presión**	pressostat de sécurité haute pression
2.8-**44**	humidistat *hygrostat* ●	**higrostato** humidostato	hygrostat *humidostat* ○
2.8-**45**	humidity controller	**dispositivo de regulación de la humedad** regulador de la humedad	régulateur d'humidité *dispositif de réglage d'humidité* ●
2.8-**46**	hunting	**oscilación periódica anormal** fluctuación	pompage
2.8-**47**	hygrostat ● *humidistat*	**higrostato** humidostato	hygrostat *humidostat* ○
2.8-**48**	inherent regulation	**autorregulación**	autorégulation
2.8-**49**	integral action controller	**regulador de acción integral**	régulateur intégral
2.8-**50**	intrinsic pressure safety	**seguridad intrínseca de la presión**	sécurité intrinsèque pour la pression
2.8-**51**	lag time *dead time*	**tiempo muerto**	temps mort
2.8-**52**	leak detection	**detección de fugas**	détection des fuites
2.8-**53**	leak detector	**detector de fugas**	détecteur de fuites
2.8-**54**	leak rate	**tasa de fugas**	débit de fuite
2.8-**55**	leakage test pressure	**presión de prueba de estanqueidad**	pression d'essai de fuite *pression de l'essai de fuite*
2.8-**56**	low-pressure controller	**presostato de baja presión**	pressostat basse pression
2.8-**57**	low-pressure safety cut-out *suction pressure safety cut-out*	**presostato de seguridad de baja presión**	pressostat de sécurité basse pression
2.8-**58**	low-suction temperature cut-out	**termostato de seguridad de aspiración**	thermostat de sécurité d'aspiration
2.8-**59**	master controller	**regulador principal**	régulateur principal

ENGLISH	ESPAÑOL	FRANÇAIS	
measuring unit	**transmisor (de medida)**	transmetteur (de mesure)	2.8-**60**
mercury switch	**interruptor de mercurio**	basculeur à mercure *interrupteur à mercure*	2.8-**61**
modulating control	**regulación modulada**	régulation modulante	2.8-**62**
monitoring	**monitorización** supervisión por señales	contrôle automatique *monitorage* ◐	2.8-**63**
oil-charging pump	**bomba de carga de aceite**	pompe de chargement d'huile	2.8-**64**
oil failure switch ○ *oil pressure cut-out* *oil pressure switch* ◐	**presostato de seguridad de aceite**	pressostat de sécurité d'huile	2.8-**65**
oil pressure cut-out *oil pressure switch* ◐ *oil failure switch* ○	**presostato de seguridad de aceite**	pressostat de sécurité d'huile	2.8-**66**
oil pressure switch ◐ *oil pressure cut-out* *oil failure switch* ○	**presostato de seguridad de aceite**	pressostat de sécurité d'huile	2.8-**67**
oil temperature cut-out	**termostato de seguridad de aceite**	thermostat de sécurité d'huile	2.8-**68**
on-off action ◐ *on-off control* *two-step control*	**regulación por todo o nada**	commande par tout ou rien *action par tout ou rien* ◐	2.8-**69**
on-off control *two-step control* *on-off action* ◐	**regulación por todo o nada**	commande par tout ou rien *action par tout ou rien* ◐	2.8-**70**
positioner	**posicionador**	positionneur	2.8-**71**
power-assisted control	**regulación indirecta**	régulation indirecte	2.8-**72**
pressostat ◐ *pressure switch* ◐	**presostato**	pressostat	2.8-**73**
pressure controller	**regulador de presión** dispositivo de regulación de presión	régulateur de pression *dispositif de réglage de pression* ◐	2.8-**74**
pressure cut-out *pressure-limiting device with manual reset* *pressure-limiting device with safety manual reset*	**limitador de presión con rearme manual** limitador de presión con rearme manual de seguridad	dispositif de limitation de pression avec réarmement manuel *dispositif de limitation de pression avec réarmement manuel de sécurité*	2.8-**75**
pressure differential cut-out	**presostato diferencial**	pressostat différentiel	2.8-**76**
pressure limiter *pressure-limiting device with automatic reset*	**limitador de presión con rearme automático**	limiteur de pression automatique *dispositif de limitation de pression avec réarmement automatique* *pressostat haute ou basse pression à réenclenchement automatique*	2.8-**77**
pressure-limiting device ◐ *pressure-relief device* ◐	**limitador de presión**	limiteur de pression	2.8-**78**
pressure-limiting device with automatic reset *pressure limiter*	**limitador de presión con rearme automático**	limiteur de pression automatique *dispositif de limitation de pression avec réarmement automatique* *pressostat haute ou basse pression à réenclenchement automatique*	2.8-**79**
pressure-limiting device with manual reset *pressure cut-out* *pressure-limiting device with safety manual reset*	**limitador de presión con rearme manual** limitador de presión con rearme manual de seguridad	dispositif de limitation de pression avec réarmement manuel *dispositif de limitation de pression avec réarmement manuel de sécurité*	2.8-**80**

	ENGLISH	ESPAÑOL	FRANÇAIS
2.8-81	pressure-limiting device with safety manual reset *pressure cut-out* *pressure-limiting device with manual reset*	**limitador de presión con rearme manual** limitador de presión con rearme manual de seguridad	dispositif de limitation de pression avec réarmement manuel *dispositif de limitation de pression avec réarmement manuel de sécurité*
2.8-82	pressure-limiting type valve	**válvula limitadora de presión**	détendeur à pression maximale d'ouverture (MOP) *détendeur à charge limitée*
2.8-83	pressure-relief device o *pressure-limiting device* o	**limitador de presión**	limiteur de pression
2.8-84	pressure-relief valve *safety valve* o	**válvula de seguridad** válvula limitadora de presión	soupape de sûreté *soupape limiteur de pression* o
2.8-85	pressure switch o *pressostat* o	**presostato**	pressostat
2.8-86	proportional-action controller	**regulador de acción proporcional**	régulateur proportionnel *régulateur à simple action*
2.8-87	Proportional Integral Derivative (PID) controller	**regulador PID (proporcional-integral-diferencial)**	régulateur PID
2.8-88	quick-closing valve	**válvula de cierre rápido**	robinet à fermeture rapide *soupape à fermeture rapide*
2.8-89	refrigerant detector	**detector de fluido frigorígeno**	détecteur de fluide frigorigène
2.8-90	relay	**relé**	relais
2.8-91	remote bulb thermostat	**termostato de bulbo**	thermostat à bulbe (et capillaire)
2.8-92	remote control	**control a distancia** mando a distancia telemando	télécommande *commande à distance* o
2.8-93	response time	**tiempo de respuesta**	temps de réponse
2.8-94	room thermostat	**termostato de ambiente**	thermostat d'ambiance
2.8-95	rupture disc *bursting disc* *safety disc* o *frangible disc* ○	**disco de seguridad** disco de ruptura	disque de rupture *disque de sûreté* *membrane d'éclatement*
2.8-96	rupture member	**dispositivo de ruptura**	dispositif de rupture *élément de rupture*
2.8-97	safety cut-out	**dispositivo de seguridad por corte**	dispositif de sécurité par coupure
2.8-98	safety device	**dispositivo de seguridad** órgano de seguridad	dispositif de sécurité *organe de sécurité*
2.8-99	safety disc o *bursting disc* *rupture disc* *frangible disc* ○	**disco de seguridad** disco de ruptura	disque de rupture *disque de sûreté* *membrane d'éclatement*
2.8-100	safety pressure cut-out	**dispositivo protector de sobrepresión**	pressostat de sécurité
2.8-101	safety valve o *pressure-relief valve*	**válvula de seguridad** válvula limitadora de presión	soupape de sûreté *soupape limiteur de pression* o
2.8-102	selector switch o *change-over switch*	**commutador**	commutateur
2.8-103	self-contained breathing apparatus	**respirador autónomo**	appareil de protection respiratoire autonome
2.8-104	self-operated control o	**regulación directa**	régulation directe
2.8-105	self-operated measuring unit o	**transmisor directo**	transmetteur direct

ENGLISH	ESPAÑOL	FRANÇAIS	
sensing bulb ◐ *thermostat bulb* *feeler (bulb)* ○ *thermostat vial* ○	**bulbo termostático** bulbo sensible	bulbe (sensible) *bulbe (thermostatique)*	2.8-**106**
sensing element ◐ *sensor* *detecting element*	**sensor** detector elemento sensible	capteur *élément sensible*	2.8-**107**
sensor *detecting element* *sensing element* ◐	**sensor** detector elemento sensible	capteur *élément sensible*	2.8-**108**
servocontrol	**servoaccionamiento**	servocommande	2.8-**109**
set point (of a controller) *set value (of a controller)*	**valor de ajuste** punto de consigna	point de consigne *point de réglage* *valeur de consigne* ○ *valeur de réglage* ○	2.8-**110**
set value (of a controller) *set point (of a controller)*	**valor de ajuste** punto de consigna	point de consigne *point de réglage* *valeur de consigne* ○ *valeur de réglage* ○	2.8-**111**
shut-off device	**dispositivo de cierre**	dispositif d'arrêt	2.8-**112**
solid-state device	**aparato transistorizado**	appareil transistorisé	2.8-**113**
spring-loaded pressure-relief valve	**válvula de seguridad de resorte**	soupape de sûreté à ressort	2.8-**114**
step control *step-by-step control* ○	**regulación paso a paso**	commande pas-à-pas	2.8-**115**
step-by-step control ○ *step control*	**regulación paso a paso**	commande pas-à-pas	2.8-**116**
step controller	**regulación paso a paso** "step controller"	"step controller" *dispositif de commande pas-à-pas* ◐ *dispositif de réglage pas-à-pas* ◐	2.8-**117**
strength-test pressure	**presión de prueba** presión de ensayo	pression de l'essai de résistance *pression d'épreuve* ◐	2.8-**118**
submaster	**controlador secundario** "submaster"	régulateur secondaire	2.8-**119**
suction pressure safety cut-out *low-pressure safety cut-out*	**presostato de seguridad de baja presión**	pressostat de sécurité basse pression	2.8-**120**
temperature controller	**dispositivo de regulación de temperatura** regulador de temperatura	régulateur de température *dispositif de réglage de température* ◐	2.8-**121**
temperature-limiting device	**dispositivo limitador de temperatura**	dispositif de limitation de la température	2.8-**122**
thermostat	**termostato**	thermostat	2.8-**123**
thermostat bulb *sensing bulb* ◐ *feeler (bulb)* ○ *thermostat vial* ○	**bulbo termostático** bulbo sensible	bulbe (sensible) *bulbe (thermostatique)*	2.8-**124**
thermostat vial ○ *thermostat bulb* *sensing bulb* ◐ *feeler (bulb)* ○	**bulbo termostático** bulbo sensible	bulbe (sensible) *bulbe (thermostatique)*	2.8-**125**
timer	**temporizador**	chronorelais *minuteur* *relais chronométrique*	2.8-**126**
transmitter	**transmisor indirecto**	transmetteur (indirect)	2.8-**127**
two-step control *on-off control* *on-off action* ◐	**regulación por todo o nada**	commande par tout ou rien *action par tout ou rien* ◐	2.8-**128**

	ENGLISH	ESPAÑOL	FRANÇAIS
	SECTION 2.9 *Working fluids* **SECTION 2.9.1** *Working fluids: general background*	**SUBCAPÍTULO 2.9** *Fluidos de trabajo* **SUBCAPÍTULO 2.9.1** *Generalidades sobre los fluidos de trabajo*	**SOUS-CHAPITRE 2.9** *Fluides actifs* **SOUS-CHAPITRE 2.9.1** *Généralités sur les fluides actifs*
2.9.1-**1**	azeotrope *azeotropic mixture*	**azeótropo** mezcla azeotrópica	azéotrope *mélange azéotropique* ◖
2.9.1-**2**	azeotropic mixture *azeotrope*	**azeótropo** mezcla azeotrópica	azéotrope *mélange azéotropique* ◖
2.9.1-**3**	blend	**mezcla**	mélange *assemblage* ◖
2.9.1-**4**	compound	**compuesto**	composé
2.9.1-**5**	compression-suction method	**método (de recuperación) aspiración-compresión**	méthode (de récupération) par aspiration-compression
2.9.1-**6**	coolant *cooling medium* *refrigerating medium*	**fluido refrigerante** fluido frigorífero	agent de refroidissement
2.9.1-**7**	cooling medium *coolant* *refrigerating medium*	**fluido refrigerante** fluido frigorífero	agent de refroidissement
2.9.1-**8**	cyclic compound	**compuesto cíclico**	composé cyclique
2.9.1-**9**	fluid quality (x)	**título del fuído (x)**	concentration massique *titre massique* ◖
2.9.1-**10**	fractionation	**destilación fraccionada**	fractionnement
2.9.1-**11**	glide *temperature glide*	**deslizamiento** deslizamiento de temperatura	glissement (de température)
2.9.1-**12**	isomer	**isómero**	isomère
2.9.1-**13**	liquid	**líquido**	liquide
2.9.1-**14**	lower flammability limit	**límite inferior de inflamabilidad**	limite inférieure d'inflammabilité
2.9.1-**15**	multi-phase fluid	**fluído multifase**	fluide polyphasique
2.9.1-**16**	natural refrigerant	**fluído frigorígeno natural**	fluide (frigorigène) naturel
2.9.1-**17**	near azeotropic	**cuasi-azeótropo**	quasi azéotropique
2.9.1-**18**	non-azeotropic mixture *zeotrope* *zeotropic mixture*	**zeótropo** mezcla zeotrópica mezcla no-azeotrópica	zéotrope *mélange non azéotropique* ◖ *mélange zéotropique* ◖
2.9.1-**19**	primary fluid *primary refrigerant*	**fluido frigorígeno primario** fluido primario	(fluide) frigorigène primaire
2.9.1-**20**	primary refrigerant *primary fluid*	**fluido frigorígeno primario** fluido primario	(fluide) frigorigène primaire
2.9.1-**21**	reclaim (to)	**regenerar**	régénérer
2.9.1-**22**	recover (to)	**recuperar**	récupérer
2.9.1-**23**	recovery equipment *refrigerant recovery unit*	**equipo para recuperación de fluido frigorígeno**	système de récupération de frigorigène
2.9.1-**24**	recycle (to)	**reciclar**	recycler
2.9.1-**25**	refrigerant	**fluido frigorífico** refrigerante (subst.)	(fluide) frigorigène

ENGLISH	ESPAÑOL	FRANÇAIS	
refrigerant recovery unit *recovery equipment*	**equipo para recuperación de fluido frigorígeno**	système de récupération de frigorigène	2.9.1-**26**
refrigerating medium *coolant* *cooling medium*	**fluido refrigerante** fluido frigorífero	agent de refroidissement	2.9.1-**27**
saturated compound	**compuesto saturado**	composé saturé	2.9.1-**28**
secondary coolant ○ *secondary refrigerant* *secondary fluid*	**fluido frigorífico secundario** refrigerante secundario fluido secundario	fluide secondaire *(fluide) frigoporteur*	2.9.1-**29**
secondary fluid *secondary refrigerant* *secondary coolant* ○	**fluido frigorífico secundario** refrigerante secundario fluido secundario	fluide secondaire *(fluide) frigoporteur*	2.9.1-**30**
secondary refrigerant *secondary fluid* *secondary coolant* ○	**fluido frigorífico secundario** refrigerante secundario fluido secundario	fluide secondaire *(fluide) frigoporteur*	2.9.1-**31**
single-phase fluid	**fluído monofásico**	fluide monophasique	2.9.1-**32**
temperature glide *glide*	**deslizamiento** deslizamiento de temperatura	glissement (de température)	2.9.1-**33**
toxicity	**toxicidad**	toxicité	2.9.1-**34**
two-phase fluid	**fluído bifásico**	fluide diphasique	2.9.1-**35**
zeotrope *non-azeotropic mixture* *zeotropic mixture*	**zeótropo** mezcla zeotrópica mezcla no-azeotrópica	zéotrope *mélange non azéotropique* ○ *mélange zéotropique* ○	2.9.1-**36**
zeotropic mixture *non-azeotropic mixture* *zeotrope*	**zeótropo** mezcla zeotrópica mezcla no-azeotrópica	zéotrope *mélange non azéotropique* ○ *mélange zéotropique* ○	2.9.1-**37**

SECTION 2.9.2 *Refrigerants* **SECTION 2.9.2.1** *HFCs*	**SUBCAPÍTULO 2.9.2** *Fluidos frigorígenos* **SUBCAPÍTULO 2.9.2.1** *HFCs*	**SOUS-CHAPITRE 2.9.2** *Fluides frigorigènes* **SOUS-CHAPITRE 2.9.2.1** *HFC*	
HFC *hydrofluorocarbon*	**hidrofluorocarburo** HFC	hydrofluorocarbure *HFC*	2.9.2.1-**1**
hydrofluorocarbon *HFC*	**hidrofluorocarburo** HFC	hydrofluorocarbure *HFC*	2.9.2.1-**2**
R125	**R125**	R125	2.9.2.1-**3**
R134a	**R134a**	R134a	2.9.2.1-**4**
R14	**R14**	R14	2.9.2.1-**5**
R143a	**R143a**	R143a	2.9.2.1-**6**
R152a	**R152a**	R152a	2.9.2.1-**7**
R218	**R218**	R218	2.9.2.1-**8**
R23	**R23**	R23	2.9.2.1-**9**
R32	**R32**	R32	2.9.2.1-**10**
RC318	**RC318**	RC318	2.9.2.1-**11**

	ENGLISH	ESPAÑOL	FRANÇAIS
	SECTION 2.9.2.2 *HCFCs*	**SUBCAPÍTULO 2.9.2.2** *HCFCs*	**SOUS-CHAPITRE 2.9.2.2** *HCFC*
2.9.2.2-**1**	HCFC *hydrochlorofluorocarbon*	**hidroclorofluorocarburo** HCFC	hydrochlorofluorocarbure *HCFC*
2.9.2.2-**2**	hydrochlorofluorocarbon *HCFC*	**hidroclorofluorocarburo** HCFC	hydrochlorofluorocarbure *HCFC*
2.9.2.2-**3**	R123	**R123**	R123
2.9.2.2-**4**	R124	**R124**	R124
2.9.2.2-**5**	R141b	**R141b**	R141b
2.9.2.2-**6**	R142b	**R142b**	R142b
2.9.2.2-**7**	R21	**R21**	R21
2.9.2.2-**8**	R22	**R22**	R22

	ENGLISH	ESPAÑOL	FRANÇAIS
	SECTION 2.9.2.3 *CFCs*	**SUBCAPÍTULO 2.9.2.3** *CFCs*	**SOUS-CHAPITRE 2.9.2.3** *CFC*
2.9.2.3-**1**	CFC *chorofluorocarbon*	**clorofluorocarburo** CFC	chlorofluorocarbure *CFC*
2.9.2.3-**2**	chorofluorocarbon *CFC*	**clorofluorocarburo** CFC	chlorofluorocarbure *CFC*
2.9.2.3-**3**	R11	**R11**	R11
2.9.2.3-**4**	R12	**R12**	R12
2.9.2.3-**5**	R13	**R13**	R13
2.9.2.3-**6**	R13B1	**R13B1**	R13B1
2.9.2.3-**7**	R113	**R113**	R113
2.9.2.3-**8**	R114	**R114**	R114
2.9.2.3-**9**	R115	**R115**	R115

	ENGLISH	ESPAÑOL	FRANÇAIS
	SECTION 2.9.2.4 *Mixtures*	**SUBCAPÍTULO 2.9.2.4** *Mezclas*	**SOUS-CHAPITRE 2.9.2.4** *Mélanges*
2.9.2.4-**1**	R401A	**R401A**	R401A
2.9.2.4-**2**	R401B	**R401B**	R401B
2.9.2.4-**3**	R401C	**R401C**	R401C
2.9.2.4-**4**	R402A	**R402A**	R402A
2.9.2.4-**5**	R402B	**R402B**	R402B
2.9.2.4-**6**	R403A	**R403A**	R403A
2.9.2.4-**7**	R403B	**R403B**	R403B
2.9.2.4-**8**	R404A	**R404A**	R404A
2.9.2.4-**9**	R405A	**R405A**	R405A

ENGLISH	ESPAÑOL	FRANÇAIS	
R406A	**R406A**	R406A	2.9.2.4-**10**
R407A	**R407A**	R407A	2.9.2.4-**11**
R407B	**R407B**	R407B	2.9.2.4-**12**
R407C	**R407C**	R407C	2.9.2.4-**13**
R407D	**R407D**	R407D	2.9.2.4-**14**
R407E	**R407E**	R407E	2.9.2.4-**15**
R408A	**R408A**	R408A	2.9.2.4-**16**
R409A	**R409A**	R409A	2.9.2.4-**17**
R409B	**R409B**	R409B	2.9.2.4-**18**
R410A	**R410A**	R410A	2.9.2.4-**19**
R410B	**R410B**	R410B	2.9.2.4-**20**
R411A	**R411A**	R411A	2.9.2.4-**21**
R411B	**R411B**	R411B	2.9.2.4-**22**
R412A	**R412A**	R412A	2.9.2.4-**23**
R413A	**R413A**	R413A	2.9.2.4-**24**
R500	**R500**	R500	2.9.2.4-**25**
R502	**R502**	R502	2.9.2.4-**26**
R503	**R503**	R503	2.9.2.4-**27**
R507A	**R507**	R507A	2.9.2.4-**28**
R508A	**R508A**	R508A	2.9.2.4-**29**
R508B	**R508B**	R508B	2.9.2.4-**30**
R509A	**R509A**	R509A	2.9.2.4-**31**

SECTION 2.9.2.5 *Natural and other refrigerants*	SUBCAPÍTULO 2.9.2.5 *Fluidos frigorígenos naturales y otros*	SOUS-CHAPITRE 2.9.2.5 *Frigorigènes naturels et autres frigorigènes*	
ammonia *R717*	**amoníaco** R717	ammoniac *R717*	2.9.2.5-**1**
ammonium hydroxide	**hidróxido amónico** hidróxido de amonio	hydroxyde d'ammonium *ammoniaque*	2.9.2.5-**2**
butane *R600*	**butano** R600	butane *R600*	2.9.2.5-**3**
carbon dioxide *R744*	**anhídrido carbónico** R744	dioxyde de carbone *R744*	2.9.2.5-**4**
ethane *R170*	**etano** R170	éthane *R170*	2.9.2.5-**5**
ethyl chloride	**cloruro de etilo**	chlorure d'éthyle	2.9.2.5-**6**
ethylene *R1150*	**etileno** R1150	éthylène *R1150*	2.9.2.5-**7**
fluorinated hydrocarbon refrigerant *fluorocarbon refrigerant*	**fluido frigorígeno halogenado derivado de hidrocarburos**	frigorigène fluorocarboné *frigorigène hydrocarbure fluoré*	2.9.2.5-**8**

	ENGLISH	ESPAÑOL	FRANÇAIS
2.9.2.5-9	fluorocarbon refrigerant *fluorinated hydrocarbon refrigerant*	**fluido frigorígeno halogenado derivado de hidrocarburos**	frigorigène fluorocarboné *frigorigène hydrocarbure fluoré*
2.9.2.5-10	halocarbon	**halocarburo**	hydrocarbure halogéné *halocarbure*
2.9.2.5-11	HBFC *hydrobromofluorocarbon*	**hidrobromofluorocarburo** HBFC	hydrobromofluorocarbure *HBFC*
2.9.2.5-12	HC *hydrocarbon*	**hidrocarburo** HC	hydrocarbure *HC*
2.9.2.5-13	HFE *hydrofluoroether*	**hidrofluoroéter** HFE	hydrofluoroéther *HFE*
2.9.2.5-14	hydrobromofluorocarbon *HBFC*	**hidrobromofluorocarburo** HBFC	hydrobromofluorocarbure *HBFC*
2.9.2.5-15	hydrocarbon *HC*	**hidrocarburo** HC	hydrocarbure *HC*
2.9.2.5-16	hydrofluoroether *HFE*	**hidrofluoroéter** HFE	hydrofluoroéther *HFE*
2.9.2.5-17	isobutane *R600a*	**isobutano** R600a	isobutane *R600a*
2.9.2.5-18	methane *R50*	**metano** R50	méthane *R50*
2.9.2.5-19	methyl chloride	**cloruro de metilo**	chlorure de méthyle
2.9.2.5-20	methylene chloride	**cloruro de metileno**	chlorure de méthylène
2.9.2.5-21	perfluorocarbon *PFC*	**perfluorocarburo** PFC	perfluorocarbure *PFC*
2.9.2.5-22	PFC *perfluorocarbon*	**perfluorocarburo** PFC	perfluorocarbure *PFC*
2.9.2.5-23	propane *R290*	**propano** R290	propane *R290*
2.9.2.5-24	propylene *R1270*	**propileno** R1270	propylène *R1270*
2.9.2.5-25	R1150 *ethylene*	**etileno** R1150	éthylène *R1150*
2.9.2.5-26	R1270 *propylene*	**propileno** R1270	propylène *R1270*
2.9.2.5-27	R170 *ethane*	**etano** R170	éthane *R170*
2.9.2.5-28	R290 *propane*	**propano** R290	propane *R290*
2.9.2.5-29	R50 *methane*	**metano** R50	méthane *R50*
2.9.2.5-30	R600 *butane*	**butano** R600	butane *R600*
2.9.2.5-31	R600a *isobutane*	**isobutano** R600a	isobutane *R600a*
2.9.2.5-32	R717 *ammonia*	**amoníaco** R717	ammoniac *R717*
2.9.2.5-33	R718 *water*	**agua** R718	eau *R718*
2.9.2.5-34	R744 *carbon dioxide*	**anhídrido carbónico** R744	dioxyde de carbone *R744*

ENGLISH	ESPAÑOL	FRANÇAIS	
R764 *surphur dioxide*	**anhídrido sulfuroso** R764	dioxyde de soufre *R764*	2.9.2.5-**35**
sulphur dioxide *R764*	**anhídrido sulfuroso** R764	dioxyde de soufre *R764*	2.9.2.5-**36**
water *R718*	**agua** R718	eau *R718*	2.9.2.5-**37**

SECTION 2.9.3 *Secondary refrigerants* **SECTION 2.9.3.1** *Brine*	**SUBCAPÍTULO 2.9.3** *Fluidos frigoríficos secundarios* **SUBCAPÍTULO 2.9.3.1** *Salmueras*	**SOUS-CHAPITRE 2.9.3** *Fluides frigoporteurs* **SOUS-CHAPITRE 2.9.3.1** *Saumure*	
agitator	**agitador**	agitateur	2.9.3.1-**1**
antifreeze agent	**anticongelante** (subst.)	antigel	2.9.3.1-**2**
aqueous solution	**solución acuosa**	solution aqueuse	2.9.3.1-**3**
brine *non-freeze liquid* *non-freezing solution*	**líquido incongelable**	saumure *liquide à bas point de congélation* *liquide "incongelable"* ◐	2.9.3.1-**4**
brine balance tank *brine expansion tank* *brine head tank* ◐	**recipiente de expansión de salmuera** tanque de expansión de salmuera	vase d'expansion (de saumure) *bac d'expansion de saumure*	2.9.3.1-**5**
brine cooler	**enfriador de salmuera**	refroidisseur de saumure	2.9.3.1-**6**
brine drum	**acumulador de salmuera**	tube accumulateur de saumure	2.9.3.1-**7**
brine expansion tank *brine balance tank* *brine head tank* ◐	**recipiente de expansión de salmuera** tanque de expansión de salmuera	vase d'expansion (de saumure) *bac d'expansion de saumure*	2.9.3.1-**8**
brine header	**colector de salmuera**	collecteur de saumure	2.9.3.1-**9**
brine head tank ◐ *brine balance tank* *brine expansion tank*	**recipiente de expansión de salmuera** tanque de expansión de salmuera	vase d'expansion (de saumure) *bac d'expansion de saumure*	2.9.3.1-**10**
brine mixing tank	**tanque de mezcla de salmuera**	bac de mélange de saumure	2.9.3.1-**11**
brine pump	**bomba de salmuera**	pompe à saumure	2.9.3.1-**12**
brine return tank	**depósito de retorno de salmuera**	réservoir à retour de saumure	2.9.3.1-**13**
brine sparge	**aspersor de salmuera** distribuidor de salmuera por aspersión	distributeur de saumure à aspersion	2.9.3.1-**14**
brine spray	**pulverización de salmuera**	pulvérisation de saumure	2.9.3.1-**15**
brine tank	**tanque de salmuera**	bac à saumure	2.9.3.1-**16**
closed-brine system	**sistema de salmuera en circuito cerrado**	système fermé à saumure	2.9.3.1-**17**
cooling mixture	**mezcla refrigerante** mezcla enfriadora	mélange refroidisseur	2.9.3.1-**18**
corrosion inhibitor	**inhibidor (de corrosión)**	inhibiteur de corrosion	2.9.3.1-**19**
densimeter *hydrometer* ◐ *salinometer* ◐ *twaddle gauge* ○	**areómetro** densímetro	densimètre *aréomètre* ○ *pèse-saumure* ○	2.9.3.1-**20**
eutectic mixture	**mezcla eutéctica**	mélange eutectique	2.9.3.1-**21**
freezant	**medio congelador**	médium congélateur	2.9.3.1-**22**

	ENGLISH	ESPAÑOL	FRANÇAIS
2.9.3.1-**23**	freezing mixture	**mezcla congeladora**	mélange congélateur
2.9.3.1-**24**	heat-transfer fluid *heat-transfer medium*	**fluido de transmisión térmica** fluido caloportador	fluide caloporteur
2.9.3.1-**25**	heat-transfer medium *heat-transfer fluid*	**fluido de transmisión térmica** fluido caloportador	fluide caloporteur
2.9.3.1-**26**	hydrometer ◐ *densimeter* *salinometer* ◐ *twaddle gauge* ○	**areómetro** densímetro	densimètre *aréomètre* ○ *pèse-saumure* ○
2.9.3.1-**27**	non-freeze liquid *brine* *non-freezing solution*	**líquido incongelable**	saumure *liquide à bas point de congélation* *liquide "incongelable"* ◐
2.9.3.1-**28**	non-freezing solution *brine* *non-freeze liquid*	**líquido incongelable**	saumure *liquide à bas point de congélation* *liquide "incongelable"* ◐
2.9.3.1-**29**	open-brine system	**sistema de salmuera en circuito abierto**	système ouvert à saumure
2.9.3.1-**30**	salinometer ◐ *densimeter* *hydrometer* ◐ *twaddle gauge* ○	**areómetro** densímetro	densimètre *aréomètre* ○ *pèse-saumure* ○
2.9.3.1-**31**	secondary loop	**circuito del refrigerante secundario**	circuit frigoporteur *circuit secondaire* ◐
2.9.3.1-**32**	twaddle gauge ○ *densimeter* *hydrometer* ◐ *salinometer* ◐	**areómetro** densímetro	densimètre *aréomètre* ○ *pèse-saumure* ○

SECTION 2.9.3.2 *Chilled and iced water*	**SUBCAPÍTULO 2.9.3.2** *Agua enfriada y agua helada*	**SOUS-CHAPITRE 2.9.3.2** *Eau réfrigérée et eau glacée*

	ENGLISH	ESPAÑOL	FRANÇAIS
2.9.3.2-**1**	chilled water	**agua enfriada** agua helada	eau réfrigérée *eau glacée* ◐
2.9.3.2-**2**	ice bank tank *ice build-up tank*	**tanque de acumulación de hielo**	bac à accumulation de glace
2.9.3.2-**3**	ice build-up tank *ice bank tank*	**tanque de acumulación de hielo**	bac à accumulation de glace
2.9.3.2-**4**	ice slurry *pumpable ice*	**hielo líquido** hielo bombeable	coulis de glace *(coulis de) glace "pompable"*
2.9.3.2-**5**	ice water	**agua de fusión del hielo** hielo fundente	eau de fusion de la glace
2.9.3.2-**6**	iced water	**agua helada**	eau glacée
2.9.3.2-**7**	iced water tank	**tanque de agua helada**	bac à eau glacée
2.9.3.2-**8**	pumpable ice *ice slurry*	**hielo líquido** hielo bombeable	coulis de glace *(coulis de) glace "pompable"*

SECTION 2.9.3.3 *Other secondary refrigerants*	**SUBCAPÍTULO 2.9.3.3** *Otros fluidos frigoríficos secundarios*	**SOUS-CHAPITRE 2.9.3.3** *Autres frigoporteurs*

	ENGLISH	ESPAÑOL	FRANÇAIS
2.9.3.3-**1**	aqua ammonia	**solución amoniacal**	liqueur ammoniacale *alcali*

ENGLISH	ESPAÑOL	FRANÇAIS	
calcium chloride	**cloruro cálcico**	chlorure de calcium	2.9.3.3-**2**
carbon dioxide snow *CO₂ snow*	**nieve carbónica**	neige carbonique	2.9.3.3-**3**
CO₂ snow *carbon dioxide snow*	**nieve carbónica**	neige carbonique	2.9.3.3-**4**
cryohydrate	**criohidrato**	cryohydrate	2.9.3.3-**5**
dry ice	**anhídrido carbónico sólido** hielo seco	glace sèche *glace carbonique*	2.9.3.3-**6**
ethanol *ethyl alcohol* ⊙	**alcohol etílico** etanol	éthanol *alcool éthylique*	2.9.3.3-**7**
ethyl alcohol ⊙ *ethanol*	**alcohol etílico** etanol	éthanol *alcool éthylique*	2.9.3.3-**8**
ethylene glycol	**etilenglicol**	éthylène glycol	2.9.3.3-**9**
eutectic ice *frozen eutectic solution*	**mezcla eutéctica** hielo eutéctico	glace eutectique *solution eutectique congelée*	2.9.3.3-**10**
frozen eutectic solution *eutectic ice*	**mezcla eutéctica** hielo eutéctico	glace eutectique *solution eutectique congelée*	2.9.3.3-**11**
glycol water	**agua glicolada**	eau glycolée	2.9.3.3-**12**
magnesium chloride	**cloruro magnésico**	chlorure de magnésium	2.9.3.3-**13**
methanol *methyl alcohol*	**alcohol metílico** metanol	méthanol *alcool méthylique*	2.9.3.3-**14**
methyl alcohol *methanol*	**alcohol metílico** metanol	méthanol *alcool méthylique*	2.9.3.3-**15**
potassium acetate	**acetato potásico**	acétate de potassium	2.9.3.3-**16**
potassium carbonate	**carbonato potásico**	carbonate de potassium	2.9.3.3-**17**
propylene glycol	**propilenglicol**	propylène glycol	2.9.3.3-**18**
sodium chloride	**cloruro sódico**	chlorure de sodium	2.9.3.3-**19**

SECTION 2.9.4 *Other working fluids*	SUBCAPÍTULO 2.9.4 *Otros fluidos de trabajo*	SOUS-CHAPITRE 2.9.4 *Autres fluides actifs*	
activated carbon	**carbón activo** carbón activado	charbon actif	2.9.4-**1**
ammonia-water	**agua amoniacal**	couple eau-ammoniac	2.9.4-**2**
lithium bromide	**bromuro de litio**	bromure de lithium	2.9.4-**3**
lithium bromide-water	**solución de bromuro de litio**	couple eau-bromure de lithium	2.9.4-**4**
silica gel	**gel de sílice**	gel de silice	2.9.4-**5**
zeolite	**zeolita** ceolita	zéolite *zéolithe*	2.9.4-**6**

Capítulo 3.

Instalaciones frigoríficas

● término aceptado

○ término obsoleto

ENGLISH	ESPAÑOL	FRANÇAIS	
SECTION 3.1 *Refrigerating equipment:* *general background*	**SUBCAPÍTULO 3.1** *Generalidades sobre instalaciones* *frigoríficas*	**SOUS-CHAPITRE 3.1** *Généralités sur les installations* *frigorifiques*	
cascade refrigerating system	**instalación frigorífica en cascada**	installation frigorifique en cascade *système cascade*	3.1-**1**
central refrigerating plant	**instalación frigorífica centralizada**	centrale frigorifique *installation centrale de froid* ◐	3.1-**2**
commercial refrigerating plant	**instalación frigorífica comercial**	installation frigorifique commerciale	3.1-**3**
engine room ◐ *machinery room* *machine room* ◐	**sala de máquinas**	salle des machines	3.1-**4**
factory-assembled system ◐ *packaged unit* *self-contained system*	**sistemo frigorífico autónomo**	installation frigorifique autonome *installation frigorifique monobloc* *système frigorifique préassemblé*	3.1-**5**
human-occupied space	**local habitado**	enceinte occupée par des personnes	3.1-**6**
industrial refrigerating plant	**instalación frigorífica industrial**	installation frigorifique industrielle	3.1-**7**
machine room ◐ *machinery room* *engine room* ◐	**sala de máquinas**	salle des machines	3.1-**8**
machinery room *engine room* ◐ *machine room* ◐	**sala de máquinas**	salle des machines	3.1-**9**
multistage refrigerating plant	**instalación frigorífica de etapas múltiples** instalación frigorífica escalonada	installation frigorifique étagée *système multi-étagé*	3.1-**10**
packaged unit *self-contained system* *factory-assembled system* ◐	**sistemo frigorífico autónomo**	installation frigorifique autonome *installation frigorifique monobloc* *système frigorifique préassemblé*	3.1-**11**
refrigerating equipment	**equipo frigorífico** componente frigorífico	composant frigorifique	3.1-**12**
refrigerating installation ◐ *refrigerating plant*	**instalación frigorífica** planta frigorífica	installation frigorifique	3.1-**13**
refrigerating loop	**circuito frigorífico**	circuit frigorifique	3.1-**14**
refrigerating plant *refrigerating installation* ◐	**instalación frigorífica** planta frigorífica	installation frigorifique	3.1-**15**
self-contained system *packaged unit* *factory-assembled system* ◐	**sistemo frigorífico autónomo**	installation frigorifique autonome *installation frigorifique monobloc* *système frigorifique préassemblé*	3.1-**16**

ENGLISH	ESPAÑOL	FRANÇAIS	
SECTION 3.2 *Construction*	**SUBCAPÍTULO 3.2** *Construcción*	**SOUS-CHAPITRE 3.2** *Construction*	
access door *access hatch* ○	**portillo de servicio**	portillon de service	3.2-**1**
access hatch ○ *access door*	**portillo de servicio**	portillon de service	3.2-**2**
air curtain	**cortina de aire**	rideau d'air	3.2-**3**
beam *girder* ○	**viga**	poutre	3.2-**4**
cargo battens ◐ *wall dunnage* ◐	**enrejado vertical**	vaigrage à claire-voie (marine)	3.2-**5**

	ENGLISH	ESPAÑOL	FRANÇAIS
3.2-6	cavity brick construction ○	**muro de bloques huecos**	maçonnerie de corps creux *parpaings creux*
3.2-7	cavity wall ○	**muro de doble pared con cámara de aire**	mur à double paroi *mur à vide d'air* ○
3.2-8	crawl space *guard space* ●	**vacío sanitario**	vide sanitaire
3.2-9	curtain wall	**muro cortina**	mur rideau
3.2-10	door gasket	**junta de puerta**	joint de porte
3.2-11	dunnage	**espaciadores para estiba** listones para estibar	lattes d'arrimage *lattis d'arrimage*
3.2-12	flexible door	**puerta flexible**	porte souple
3.2-13	floor dunnage ● *floor rack* *floor grating* ●	**enrejado**	caillebotis
3.2-14	floor grating ● *floor rack* *floor dunnage* ●	**enrejado**	caillebotis
3.2-15	floor rack *floor dunnage* ● *floor grating* ●	**enrejado**	caillebotis
3.2-16	flush fitting door *infitting door* ●	**puerta empotrada**	porte encastrée
3.2-17	frame wall ●	**muro de estructura** muro de carga	mur à ossature
3.2-18	framework	**estructura**	charpente
3.2-19	frost heave ●	**levantamiento del suelo por congelación**	soulèvement du sol par congélation
3.2-20	girder ○ *beam*	**viga**	poutre
3.2-21	guard space ● *crawl space*	**vacío sanitario**	vide sanitaire
3.2-22	hard covering *wearing surface (of a floor)* ●	**capa de rodadura** capa de desgaste	chape d'usure
3.2-23	heater cable ● *heater strip* *heater tape* ○	**cordón calefactor** hilo calefactor	bande chauffante *câble chauffant* *cordon chauffant*
3.2-24	heater mat	**capa calefactora**	nappe chauffante *réseau antigel*
3.2-25	heater strip *heater cable* ● *heater tape* ○	**cordón calefactor** hilo calefactor	bande chauffante *câble chauffant* *cordon chauffant*
3.2-26	heater tape ○ *heater strip* *heater cable* ●	**cordón calefactor** hilo calefactor	bande chauffante *câble chauffant* *cordon chauffant*
3.2-27	hung ceiling ● *suspended ceiling*	**techo suspendido**	plafond suspendu
3.2-28	infitting door ● *flush fitting door*	**puerta empotrada**	porte encastrée
3.2-29	inspection window *porthole* ●	**ventanillo de inspección** mirilla de inspección	hublot
3.2-30	insulated door	**puerta aislada**	porte isolante *porte isolée*

ENGLISH	ESPAÑOL	FRANÇAIS	
insulated openings	**carpintería aislante**	menuiseries isolantes *menuiseries isothermes* ○	3.2-**31**
insulated web (of a girder) ○	**alma aislante (de viga)**	âme isolante (de poutre)	3.2-**32**
masonry wall	**muro de mamposteria**	mur en maçonnerie	3.2-**33**
mushroom floor	**forjado sobre columnas fungiformes** forjado sobre capiteles	plancher champignon	3.2-**34**
overhead rail	**raíl aéreo**	rail aérien	3.2-**35**
overlap door	**puerta superpuesta**	porte en surépaisseur *porte en applique* ○	3.2-**36**
panelling	**revestimiento de protección**	bardage *parement*	3.2-**37**
peep-hole	**mirilla**	judas	3.2-**38**
plug door	**portillo de visita**	tampon de visite	3.2-**39**
porthole ○ *inspection window*	**ventanillo de inspección** mirilla de inspección	hublot	3.2-**40**
powered door	**puerta automática**	porte commandée	3.2-**41**
protective coating	**recubrimiento protector**	revêtement de protection	3.2-**42**
purlin ○	**correa**	panne	3.2-**43**
rafter ○	**cabio**	chevron	3.2-**44**
sliding door	**puerta corredera**	porte coulissante	3.2-**45**
soil freezing	**congelación del suelo**	congélation du sol	3.2-**46**
suspended ceiling *hung ceiling* ○	**techo suspendido**	plafond suspendu	3.2-**47**
swinging door	**puerta pivotante**	porte battante *porte va-et-vient*	3.2-**48**
truss (of a roof) ○	**cercha**	ferme	3.2-**49**
underfloor ventilation	**ventilación por falso techo**	ventilation sous plancher *plancher soufflant*	3.2-**50**
ventilated crawl space	**vacío sanitario ventilado**	vide sanitaire ventilé	3.2-**51**
wall dunnage ○ *cargo battens* ○	**enrejado vertical**	vaigrage à claire-voie (marine)	3.2-**52**
wall rail	**pequeño perfil de atado horizontal**	lisse	3.2-**53**
wearing surface (of a floor) ○ *hard covering*	**capa de rodadura** capa de desgaste	chape d'usure	3.2-**54**

SECTION 3.3 *Thermal insulation* SECTION 3.3.1 *Thermal insulation: general background*	SUBCAPÍTULO 3.3 *Aislamiento térmico* SUBCAPÍTULO 3.3.1 *Generalidades sobre aislamiento térmico*	SOUS-CHAPITRE 3.3 *Isolation thermique* SOUS-CHAPITRE 3.3.1 *Généralités sur l'isolation thermique*	
backing insulation	**aislamiento protegido**	isolation protégée	3.3.1-**1**
blanket-type insulant	**manta aislante** aislante en mantas flexible	isolant en matelas souple	3.3.1-**2**

	ENGLISH	ESPAÑOL	FRANÇAIS
3.3.1-**3**	block-type insulant	**aislante en bloques**	isolant en blocs
3.3.1-**4**	board-type insulant ◐ *slab insulant* ◐	**aislante en paneles** aislante en planchas	isolant en panneau *isolant en plaque*
3.3.1-**5**	composite insulation	**aislamiento compuesto** aislamiento multicapa	isolation composite
3.3.1-**6**	conventional insulation	**aislamiento tradicional**	isolation traditionnelle
3.3.1-**7**	evacuated insulation ◐ *vacuum insulation*	**aislamiento bajo vacío**	isolation sous vide
3.3.1-**8**	fill insulation *loose-fill-type insulant* ○	**aislante de relleno**	isolant de bourrage *isolant en vrac* *isolant de remplissage*
3.3.1-**9**	foil insulant	**aislante en láminas** aislante en capas múltiples	isolant en feuilles
3.3.1-**10**	heat bridge *heat channel* ◐	**puente térmico**	pont thermique
3.3.1-**11**	heat channel ◐ *heat bridge*	**puente térmico**	pont thermique
3.3.1-**12**	heat leakage	**pérdidas de calor** fuga térmica	fuite thermique *déperdition de chaleur* ◐
3.3.1-**13**	insulate (to)	**aislar** calorifugar	calorifuger *isoler*
3.3.1-**14**	insulated	**aislado** calorifugado (adj.)	isolé
3.3.1-**15**	insulating jacket	**recubrimiento aislante**	enveloppe isolante
3.3.1-**16**	jacket	**doble pared**	double paroi
3.3.1-**17**	loose-fill-type insulant ○ *fill insulation*	**aislante de relleno**	isolant de bourrage *isolant en vrac* *isolant de remplissage*
3.3.1-**18**	mat-type insulant	**manta aislante** aislante en mantas flexible	isolant en matelas
3.3.1-**19**	powdered insulant	**aislante en polvo** aislante granulado	isolant en poudre
3.3.1-**20**	pre-formed insulation	**aislamiento pre-formado**	isolant préformé
3.3.1-**21**	slab insulant ◐ *board-type insulant* ◐	**aislante en paneles** aislante en planchas	isolant en panneau *isolant en plaque*
3.3.1-**22**	superinsulation	**aislamiento reforzado**	superisolation
3.3.1-**23**	thermal inertia	**inercia térmica**	inertie thermique
3.3.1-**24**	(thermal) insulation	**aislamiento térmico** calorifugado (subst.)	calorifugeage *isolation (thermique)* ◐
3.3.1-**25**	thermal insulation	**aislamiento térmico** calorifugado (subst.)	isolation thermique
3.3.1-**26**	thermal insulation composite system	**sistema de aislamiento térmico compuesto**	système d'isolation thermique composite
3.3.1-**27**	unbound insulation	**aislamiento sin aglutinante**	isolant non encollé
3.3.1-**28**	vacuum insulation *evacuated insulation* ◐	**aislamiento bajo vacío**	isolation sous vide
3.3.1-**29**	wall losses	**pérdidas por las paredes** pérdidas por las superficies de transmisíon	pertes (thermiques) par les parois

ENGLISH	ESPAÑOL	FRANÇAIS	
SECTION 3.3.2 *Insulating materials*	**SUBCAPÍTULO 3.3.2** *Materiales aislantes*	**SOUS-CHAPITRE 3.3.2** *matériaux isolants*	
aluminium foil	**lamina de aluminio** hoja de aluminio	feuille d'aluminium	3.3.2-**1**
blanket *mat*	**manta** fieltro mat	feutre	3.3.2-**2**
blowing wool	**lana proyectada** borra proyectada	isolant particulaire insufflable	3.3.2-**3**
bonding	**aglutinación**	ensimage	3.3.2-**4**
calcium silicate	**silicato cálcico**	silicate de calcium	3.3.2-**5**
carbon fibre	**fibra de carbono**	fibre de carbone	3.3.2-**6**
cellular concrete ○ *foam concrete* *foamed concrete*	**hormigón celular**	béton cellulaire	3.3.2-**7**
cellular glass *glass foam*	**vidrio celular**	verre cellulaire *verre mousse*	3.3.2-**8**
cellular insulant ○	**aislante celular**	isolant cellulaire	3.3.2-**9**
cellular material	**material celular**	matériau alvéolaire	3.3.2-**10**
cellular plastic *plastic foam* ○	**plástico celular** espuma plástica	mousse plastique *plastique alvéolaire* *plastique cellulaire*	3.3.2-**11**
cellular rubber ○	**caucho celular**	caoutchouc cellulaire *caoutchouc mousse*	3.3.2-**12**
cellulose insulation	**aislamiento celulósico**	isolant cellulosique	3.3.2-**13**
ceramic fibre	**fibra cerámica**	fibre céramique	3.3.2-**14**
closed-cell foamed plastic	**plástico celular de celdillas cerradas**	plastique cellulaire à cellules fermées	3.3.2-**15**
composite insulation product	**producto aislante compuesto** producto aislante multicapa	produit isolant composite	3.3.2-**16**
composite panel	**panel compuesto**	panneau composite	3.3.2-**17**
cork	**corcho**	liège	3.3.2-**18**
cork board	**plancha de corcho**	panneau de liège	3.3.2-**19**
diatomaceous earth ○	**tierra de diatomeas**	terre à diatomées	3.3.2-**20**
elastomer	**elastómero**	élastomère	3.3.2-**21**
evacuated powder	**polvo bajo vacío**	poudre sous vide	3.3.2-**22**
expanded (cellular) plastic	**plástico (celular) expandido**	plastique (cellulaire) expansé	3.3.2-**23**
expanded clay	**arcilla expandida**	argile expansée	3.3.2-**24**
expanded cork	**corcho expandido**	liège expansé	3.3.2-**25**
expanded perlite board	**plancha de perlita expandida**	panneau de perlite expansée	3.3.2-**26**
expanded polystyrene	**poliestireno expandido**	polystyrène expansé	3.3.2-**27**
expanded polyvinyl chloride	**cloruro de polivinilo expandido**	mousse de PVC	3.3.2-**28**
extruded (cellular) plastic	**plástico (celular) extruido**	plastique (cellulaire) extrudé	3.3.2-**29**
extruded polystyrene foam	**espuma de poliestireno extruido**	mousse de polystyrène extrudé	3.3.2-**30**
fibrous insulant	**aislante fibroso**	isolant fibreux	3.3.2-**31**

ENGLISH	ESPAÑOL	FRANÇAIS
3.3.2-**32** fibrous insulation	**aislamlento fibroso**	isolation fibreuse
3.3.2-**33** flexible elastomeric foam	**espuma elastomérica flexible**	mousse souple élastomère
3.3.2-**34** foam concrete *foamed concrete* *cellular concrete* ◑	**hormigón celular**	béton cellulaire
3.3.2-**35** foamed concrete *foam concrete* *cellular concrete* ◑	**hormigón celular**	béton cellulaire
3.3.2-**36** gas space	**cámara de aire** cámara de gas	lame d'air
3.3.2-**37** glass fibre *glass wool*	**fibra de vidrio** lana de vidrio	fibre de verre *laine de verre* ◑
3.3.2-**38** glass foam *cellular glass*	**vidrio celular**	verre cellulaire *verre mousse*
3.3.2-**39** glass wool *glass fibre*	**fibra de vidrio** lana de vidrio	fibre de verre *laine de verre* ◑
3.3.2-**40** granulated cork	**corcho granulado**	granulés de liège *liège granulé*
3.3.2-**41** granulated wool	**lana granulada**	laine en flocons ou en nodules
3.3.2-**42** graphite fibre	**fibra de grafito**	fibre de graphite
3.3.2-**43** in situ thermal insulation product	**aislamiento térmico producido in situ**	produit d'isolation thermique in situ
3.3.2-**44** insulant ◑ *insulating material*	**aislante** (subst.) material aislante	isolant (subst.) *matériau isolant*
3.3.2-**45** insulating ◑	**aislante** (adj.)	isolant (adj.)
3.3.2-**46** insulating castable refractory	**pieza refractaria aislante** hormigón aislante refractario	béton réfractaire isolant
3.3.2-**47** insulating concrete *lightweight concrete*	**hormigón aislante** hormigón ligero	béton allégé *béton de granulats légers* *béton isolant*
3.3.2-**48** insulating material *insulant* ◑	**aislante** (subst.) material aislante	isolant (subst.) *matériau isolant*
3.3.2-**49** insulating plaster	**escayola aislante**	plâtre isolant
3.3.2-**50** insulating rope	**cuerda aislante**	bourrelet isolant
3.3.2-**51** lightweight aggregate	**agregado ligero**	granulat léger
3.3.2-**52** lightweight concrete *insulating concrete*	**hormigón aislante** hormigón ligero	béton allégé *béton de granulats légers* *béton isolant*
3.3.2-**53** magnesia	**magnesia**	magnésie
3.3.2-**54** man-made mineral fibre	**fibra mineral artificial**	fibre minérale manufacturée
3.3.2-**55** mat *blanket*	**manta** fieltro mat	feutre
3.3.2-**56** microporous insulation	**aislante microporoso**	isolant microporeux *aérogel de silice*
3.3.2-**57** mineral fibre	**fibra mineral**	fibre minérale
3.3.2-**58** mineral wool *slag wool* ◑	**lana mineral** lana de escorias	laine de laitier *laine minérale*
3.3.2-**59** multicellular metal foil	**lámina metálica de estructura celular**	feuille métallique à structure cellulaire

ENGLISH	ESPAÑOL	FRANÇAIS	
multilayer insulant	**aislante en capas múltiples** aislante multicapa	isolant multicouche	3.3.2-**60**
opacified silica-aerogel	**aerogel de sílice opacificado** aerogel de sílice impermeabilizado	aérosilicagel opacifié	3.3.2-**61**
open-cell foamed plastic	**plástico celular de celdillas abiertas**	plastique cellulaire à cellules ouvertes	3.3.2-**62**
perlite	**perlita**	perlite	3.3.2-**63**
perlite plaster	**escayola de perlita**	plâtre de perlite	3.3.2-**64**
phenolic foam	**espuma fenólica**	mousse phénolique	3.3.2-**65**
pipe insulation	**coquilla aislante**	isolation de tuyauterie	3.3.2-**66**
plastic foam ◐ *cellular plastic*	**plástico celular** espuma plástica	mousse plastique *plastique alvéolaire* *plastique cellulaire*	3.3.2-**67**
polyethylene foam	**espuma de polietileno**	mousse de polyéthylène	3.3.2-**68**
polyisocyanurate foam	**espuma de poliisocianurato**	mousse polyisocyanurate	3.3.2-**69**
polyurethane foam	**espuma de poliuretano**	mousse rigide de polyuréthane	3.3.2-**70**
pouring wool	**lana a granel**	laine à déverser	3.3.2-**71**
prefabricated panel	**panel (aislante) prefabricado**	panneau préfabriqué	3.3.2-**72**
reflective insulant	**aislante reflectante**	isolant réfléchissant	3.3.2-**73**
reflective insulation	**aislante reflectante**	isolation réfléchissante	3.3.2-**74**
rock wool	**lana de roca**	laine de roche	3.3.2-**75**
sandwich panel	**panel tipo sandwich** panel sandwich	panneau sandwich	3.3.2-**76**
sandwich panel insulation	**panel aislante tipo sandwich**	isolation en panneaux	3.3.2-**77**
slag wool ◐ *mineral wool*	**lana mineral** lana de escorias	laine de laitier *laine minérale*	3.3.2-**78**
slotted slab	**plancha acanalada**	panneau rainuré	3.3.2-**79**
thermal insulation material	**material aislante térmico**	matériau d'isolation thermique	3.3.2-**80**
thermal insulation product	**producto aislante térmico**	produit d'isolation thermique	3.3.2-**81**
urea formaldehyde foam	**espuma de urea-formol**	mousse urée-formaldéhyde	3.3.2-**82**
vacuum insulation jacket	**aislamiento de doble pared con vacío**	enveloppe isolante sous vide	3.3.2-**83**
vacuum reflective insulation	**aislante reflectante al vacio** aislante especular al vacio	isolation réfléchissante sous vide	3.3.2-**84**
vermiculite	**vermiculita**	vermiculite	3.3.2-**85**
wood wool	**lana de madera**	laine de bois	3.3.2-**86**
wood wool board	**tablero aglomerado de viruta de madera**	panneau en laine de bois	3.3.2-**87**

SECTION 3.3.3 *Installation of insulants*	SUBCAPÍTULO 3.3.3 *Colocación de materiales aislantes*	SOUS-CHAPITRE 3.3.3 *Installation des matériaux isolants*	
binder	**aglutinante**	liant	3.3.3-**1**
block	**bloque**	bloc isolant	3.3.3-**2**
blown insulation	**aislante proyectado**	isolant par soufflage	3.3.3-**3**

	ENGLISH	ESPAÑOL	FRANÇAIS
3.3.3-**4**	board *slab*	**plancha**	panneau isolant
3.3.3-**5**	breaker strip ◐	**varilla de fijación aislante**	entretoise isolante *barrette de maintien* ◐ *rupteur de conduction* ◐
3.3.3-**6**	cladding	**revestimiento laminar**	revêtement
3.3.3-**7**	coating	**capa de revestimiento**	enduit de finition
3.3.3-**8**	curved board	**tablero curvo**	panneau incurvé
3.3.3-**9**	elbow	**codo**	coude
3.3.3-**10**	embedded insulation	**aislamiento embebido**	isolation en fond de coffrage
3.3.3-**11**	facing	**revestimiento**	parement
3.3.3-**12**	finishing cement	**cemento de acabado**	enduit de finition *enduit de finition (ciment)* *enduit hydraulique de finition*
3.3.3-**13**	foamed in-place insulation *foamed in-situ insulation* ◐	**aislamiento expandido in situ**	isolation expansée in situ
3.3.3-**14**	foamed in-situ insulation ◐ *foamed in-place insulation*	**aislamiento expandido in situ**	isolation expansée in situ
3.3.3-**15**	insulated suspending tiebar	**barra de suspensión aislante**	suspente isolante
3.3.3-**16**	insulating brick	**bloque aislante**	brique isolante
3.3.3-**17**	insulating cement	**cemento aislante**	enduit isolant *ciment isolant*
3.3.3-**18**	insulating joint	**junta de compresión aislante**	joint de retrait isolant
3.3.3-**19**	insulating mastic	**masilla aislante**	mastic isolant
3.3.3-**20**	insulation cover	**coquilla aislante**	coquille isolante
3.3.3-**21**	insulation finish	**acabado del aislante**	enduit pour isolation
3.3.3-**22**	laminate	**laminado**	produit feuilleté *produit laminé* ◐
3.3.3-**23**	mattress	**colchón**	matelas isolant
3.3.3-**24**	mitred joint	**unión en inglete**	onglet
3.3.3-**25**	moulding	**moldura**	bandelette isolante
3.3.3-**26**	pipe section	**porción de tuberia**	fourreau isolant
3.3.3-**27**	pneumatic application	**aplicación neumática**	application pneumatique
3.3.3-**28**	poured application	**aplicación por vertido**	application par déversement *application par remplissage*
3.3.3-**29**	radiation shield	**protección contra la radiación**	écran antirayonnement
3.3.3-**30**	roll	**rollo**	rouleau
3.3.3-**31**	self-supporting insulation	**aislamiento autoportante**	isolation autoportante
3.3.3-**32**	slab *board*	**plancha**	panneau isolant
3.3.3-**33**	sprayed insulation ◐ *sprayed-on insulation* ◐	**aislamiento por proyección**	isolation par projection
3.3.3-**34**	sprayed-on insulation ◐ *sprayed insulation* ◐	**aislamiento por proyección**	isolation par projection
3.3.3-**35**	structural insulation	**aislamiento soporte**	isolation porteuse

ENGLISH	ESPAÑOL	FRANÇAIS	
tube insulation	**coquilla aislante**	fourreau isolant *manchon isolant*	3.3.3-**36**
wire mat	**manta con armadura**	nappe grillagée	3.3.3-**37**

SECTION 3.3.4 *Gas and vapour seals*	SUBCAPÍTULO 3.3.4 *Estanqueidad al vapor y a los gases*	SOUS-CHAPITRE 3.3.4 *Etanchéité à la vapeur et aux gaz*	
accumulation test	**ensayo por acumulación**	contrôle par accumulation	3.3.4-**1**
airtight	**estanco al aire** hermético al aire	étanche à l'air	3.3.4-**2**
asphalted paper	**papel asfaltado**	papier bitumé	3.3.4-**3**
bituminous felt	**fieltro bituminoso** fieltro asfaltado	feutre bitumineux	3.3.4-**4**
breather plug *vent* ○	**ventanillo de aireación** respiradero	évent	3.3.4-**5**
calibration leak	**calibrado de fuga**	fuite calibrée	3.3.4-**6**
capillary leak	**fuga capilar**	fuite capillaire	3.3.4-**7**
coating barrier ○	**recubrimiento de estanqueidad**	endult d'étanchéité	3.3.4-**8**
damp proofing ○ *moisture proofing*	**impermeabilización**	imperméabilisation	3.3.4-**9**
gas-proof *gas-tight*	**estanco a los gases** hermético a los gases	étanche aux gaz	3.3.4-**10**
gas seal	**pantalla de estanqueidad a los gases** pantalla hermética a los gases	écran d'étanchéité aux gaz	3.3.4-**11**
gas-tight *gas-proof*	**estanco a los gases** hermético a los gases	étanche aux gaz	3.3.4-**12**
laminated paper	**papel estratificado**	papier stratifié	3.3.4-**13**
leak-free ○ *leaktight*	**estanco** impermeable	étanche *imperméable*	3.3.4-**14**
leakage rate	**caudal de fuga**	flux de fuite	3.3.4-**15**
leaktight *leak-free* ○	**estanco** impermeable	étanche *imperméable*	3.3.4-**16**
membrane barrier	**lámina de estanqueidad** hoja de estanqueidad	feuille d'étanchéité *membrane d'étanchéité*	3.3.4-**17**
moisture barrier *vapour barrier* *vapour seal* *water vapour barrier*	**barrera antivapor** pantalla hidrófuga barrera anti vapor de agua	barrière anti-vapeur *écran d'étanchéité à la vapeur (d'eau)* *écran pare-vapeur*	3.3.4-**18**
moisture proofing *damp proofing* ○	**impermeabilización**	imperméabilisation	3.3.4-**19**
permeability	**permeabilidad**	perméabilité	3.3.4-**20**
permeability coefficient	**coeficiente de permeabilidad**	coefficient de perméabilité	3.3.4-**21**
permeance	**permeancia**	perméance	3.3.4-**22**
plastic film	**película plástica**	pellicule plastique	3.3.4-**23**
sealing mastic *sealing putty* *sealing stopper* ○	**masilla de estanqueidad**	mastic d'étanchéité	3.3.4-**24**

ENGLISH	ESPAÑOL	FRANÇAIS
3.3.4-25 sealing putty *sealing mastic* *sealing stopper* ○	**masilla de estanqueidad**	mastic d'étanchéité
3.3.4-26 sealing stopper ○ *sealing mastic* *sealing putty*	**masilla de estanqueidad**	mastic d'étanchéité
3.3.4-27 structural barrier	**estanqueidad estructural**	étanchéité structurale
3.3.4-28 vapour barrier *moisture barrier* *vapour seal* *water vapour barrier*	**barrera antivapor** pantalla hidrófuga barrera anti vapor de agua	barrière anti-vapeur *écran d'étanchéité à la vapeur (d'eau)* *écran pare-vapeur*
3.3.4-29 vapour migration *vapour transmission* *vapour transfer* ○	**transmisión de vapor (de agua)** difusión de vapor (de agua)	migration de la vapeur (d'eau) *passage de la vapeur (d'eau)* *transfert de la vapeur (d'eau)*
3.3.4-30 vapour permeability	**permeabilidad al vapor de agua**	perméabilité à la vapeur (d'eau)
3.3.4-31 vapour-proof ○ *vapour-tight*	**estanco al vapor de agua** impermeable al vapor de agua	étanche à la vapeur (d'eau) *imperméable à la vapeur (d'eau)*
3.3.4-32 vapour seal *moisture barrier* *vapour barrier* *water vapour barrier*	**barrera antivapor** pantalla hidrófuga barrera anti vapor de agua	barrière anti-vapeur *écran d'étanchéité à la vapeur (d'eau)* *écran pare-vapeur*
3.3.4-33 vapour-tight *vapour-proof* ○	**estanco al vapor de agua** impermeable al vapor de agua	étanche à la vapeur (d'eau) *imperméable à la vapeur (d'eau)*
3.3.4-34 vapour transfer ○ *vapour migration* *vapour transmission*	**transmisión de vapor (de agua)** difusión de vapor (de agua)	migration de la vapeur (d'eau) *passage de la vapeur (d'eau)* *transfert de la vapeur (d'eau)*
3.3.4-35 vapour transmission *vapour migration* *vapour transfer* ○	**transmisión de vapor (de agua)** difusión de vapor (de agua)	migration de la vapeur (d'eau) *passage de la vapeur (d'eau)* *transfert de la vapeur (d'eau)*
3.3.4-36 vent ○ *breather plug*	**ventanillo de aireación** respiradero	évent
3.3.4-37 water vapour barrier *moisture barrier* *vapour barrier* *vapour seal*	**barrera antivapor** pantalla hidrófuga barrera anti vapor de agua	barrière anti-vapeur *écran d'étanchéité à la vapeur (d'eau)* *écran pare-vapeur*
3.3.4-38 water vapour retarder	**retardador de la difusión de vapor de agua**	pare-vapeur

	SECTION 3.4 *Sound insulation*	SUBCAPÍTULO 3.4 *Aislamiento acústico*	SOUS-CHAPITRE 3.4 *Isolation phonique*
3.4-1	acoustic insulation *sound insulation* ○	**aislamiento acústico**	isolation acoustique *isolation phonique*
3.4-2	damping (of vibration)	**amortiguación (de vibraciones)** amortiguamiento (de vibraciones)	amortissement (de vibration)
3.4-3	muffler (USA) ○ *silencer* *sound attenuator* *noise damper* ○ *sound absorber* ○	**silenciador**	silencieux *insonorisateur*
3.4-4	noise criteria curves	**curvas de ruido**	courbes de bruit

ENGLISH	ESPAÑOL	FRANÇAIS	
noise damper ○ *silencer* *sound attenuator* *muffler (USA)* ○ *sound absorber* ○	**silenciador**	silencieux *insonorisateur*	3.4-**5**
noise reduction *sound attenuation* *sound damping* ○ *sound deadening* ○	**insonorización**	insonorisation *amortissement du son* *atténuation du bruit*	3.4-**6**
silencer *sound attenuator* *muffler (USA)* ○ *noise damper* ○ *sound absorber* ○	**silenciador**	silencieux *insonorisateur*	3.4-**7**
sound absorber ○ *silencer* *sound attenuator* *muffler (USA)* ○ *noise damper* ○	**silenciador**	silencieux *insonorisateur*	3.4-**8**
sound attenuation *noise reduction* *sound damping* ○ *sound deadening* ○	**insonorización**	insonorisation *amortissement du son* *atténuation du bruit*	3.4-**9**
sound attenuator *silencer* *muffler (USA)* ○ *noise damper* ○ *sound absorber* ○	**silenciador**	silencieux *insonorisateur*	3.4-**10**
sound damping ○ *noise reduction* *sound attenuation* *sound deadening* ○	**insonorización**	insonorisation *amortissement du son* *atténuation du bruit*	3.4-**11**
sound deadening ○ *noise reduction* *sound attenuation* *sound damping* ○	**insonorización**	insonorisation *amortissement du son* *atténuation du bruit*	3.4-**12**
sound insulation ○ *acoustic insulation*	**aislamiento acústico**	isolation acoustique *isolation phonique*	3.4-**13**
sound level	**nivel sonoro**	niveau sonore	3.4-**14**
sound power level (L_W)	**nivel de potencia sonora (L_W)**	niveau de puissance acoustique (L_W)	3.4-**15**
sound pressure level	**nivel de presión sonora** nivel de presión de sonido	niveau de pression acoustique	3.4-**16**
sound trap	**trampa de ruidos** eliminador de ruidos	piège à sons	3.4-**17**

SECTION 3.5 *Operation* **SECTION 3.5.1** *Running of refrigerating equipment*	**SUBCAPÍTULO 3.5** *Explotación* **SUBCAPÍTULO 3.5.1** *Funcionamiento de las instalaciones frigoríficas*	**SOUS-CHAPITRE 3.5** *Exploitation* **SOUS-CHAPITRE 3.5.1** *Fonctionnement des installations frigorifiques*	
actuate (to) ○	**accionar** actuar	actionner ○	3.5.1-**1**
automatic starting system	**sistema de arranque automático**	appareillage de démarrage automatique *circuit de démarrage automatique* ○	3.5.1-**2**

	ENGLISH	ESPAÑOL	FRANÇAIS
3.5.1-**3**	balance pressure	**presión de equilibrio**	pression d'équilibre
3.5.1-**4**	belt drive	**accionamiento por correa** arrastre por correa	entraînement par courroie
3.5.1-**5**	breakdown (of a machine)	**avería**	panne (de machine)
3.5.1-**6**	burn-out (of a motor)	**quemado (de un motor)**	grillage (d'un moteur)
3.5.1-**7**	calculation temperature	**temperatura de cálculo**	température de calcul
3.5.1-**8**	capacity control *capacity modulation*	**regulación de la potencia (frigorífica)** variación de la potencia (de un compresor)	régulation de la puissance (frigorifique) *variation de la puissance (d'un compresseur)*
3.5.1-**9**	capacity controller *capacity regulator* ○	**regulador de potencia** dispositivo de variación de potencia	dispositif de variation de puissance *régulateur de puissance*
3.5.1-**10**	capacity modulation *capacity control*	**regulación de la potencia (frigorífica)** variación de la potencia (de un compresor)	régulation de la puissance (frigorifique) *variation de la puissance (d'un compresseur)*
3.5.1-**11**	capacity reducer ○	**reductor de potencia**	réducteur de puissance
3.5.1-**12**	capacity regulator ○ *capacity controller*	**regulador de potencia** dispositivo de variación de potencia	dispositif de variation de puissance *régulateur de puissance*
3.5.1-**13**	cavitation	**cavitación**	cavitation
3.5.1-**14**	charging	**carga (de un circuito)**	charge (d'un circuit)
3.5.1-**15**	commissioning	**puesta en marcha** puesta en servicio	mise en (état de) fonctionnement
3.5.1-**16**	control system	**sistema de control**	système de régulation
3.5.1-**17**	corrosion	**corrosión**	corrosion
3.5.1-**18**	degassing ○ *gas purging* *purging* *gas-off* ○	**purga de gas** purga	dégazage *purge*
3.5.1-**19**	design pressure maximum allowable pressure *maximum declared pressure* *maximum working pressure (MWP)*	**presión máxima de servicio** presión de diseño presión máxima admisible presión máxima declarada presión máxima de trabajo	pression maximale de service
3.5.1-**20**	design temperature	**temperatura de diseño**	température de conception spécifiée
3.5.1-**21**	direct drive	**accionamiento por acoplamiento directo** arrastre por acoplamiento directo	(entraînement par) accouplement direct
3.5.1-**22**	drive	**accionamiento** arrastre	entraînement
3.5.1-**23**	drive-through a gearbox	**transmisión por caja de cambios**	entraînement par boîte de vitesse *entraînement par une pignonnerie*
3.5.1-**24**	flash gas	**vapor instantáneo**	vapeur "instantanée" *vaporisat* ○
3.5.1-**25**	flash vaporization *instantaneous vaporization* ○	**vaporización instantánea**	vaporisation instantanée
3.5.1-**26**	flush (to)	**limpiar a presión**	faire une chasse
3.5.1-**27**	flywheel	**volante**	volant
3.5.1-**28**	freeze-up	**bloqueo por congelación** obturación por congelación	bouchage par congélation
3.5.1-**29**	gas bottle (1) ○ *refrigerant cylinder*	**botella (de suministro) de fluido frigorífico** botela de fluido frigorígeno botella de gas	bouteille de frigorigène (de livraison)

ENGLISH	ESPAÑOL	FRANÇAIS	
gas bottle (2) ○ *service cylinder* ○	**botella de fluido frigorífico (para mantenimiento)**	bouteille de frigorigène (de dépanneur)	3.5.1-**30**
gas lock *vapour lock*	**atasco por vapor** oclusión por vapor	bouchon de vapeur	3.5.1-**31**
gas-off ○ *gas purging* *purging* *degassing* ○	**purga de gas** purga	dégazage *purge*	3.5.1-**32**
gas purging *purging* *degassing* ○ *gas-off* ○	**purga de gas** *purga*	dégazage *purge*	3.5.1-**33**
gas shortage ○ *lack of refrigerant* *undercharge*	**falta de fluido**	manque de fluide	3.5.1-**34**
gauge (1)	**manómetro**	jauge	3.5.1-**35**
gauge (2)	**calibre** galga	calibre	3.5.1-**36**
hammering *liquid hammer* *pipe hammer* ○	**golpe de ariete**	coup de bélier	3.5.1-**37**
hammering (of an expansion valve) ○ *needle hammer*	**martilleo (de una válvula de expansión)** golpeteo (de une válvula de expansión)	martèlement d'un obturateur	3.5.1-**38**
holding charge *service charge* ○	**carga de servicio**	charge d'attente	3.5.1-**39**
hunting (of a valve)	**bombeo (de una válvula)** fluctuación (de una válvula)	pompage (d'un régleur)	3.5.1-**40**
initial charge	**carga inicial**	charge initiale *première charge*	3.5.1-**41**
instantaneous vaporization ○ *flash vaporization*	**vaporización instantánea**	vaporisation instantanée	3.5.1-**42**
inverter drive	**arrastre por motor con variador de frecuencia** inverter	entraînement des moteurs par variateur de fréquence	3.5.1-**43**
lack of refrigerant *undercharge* *gas shortage* ○	**falta de fluido**	manque de fluide	3.5.1-**44**
leak	**fuga**	fuite	3.5.1-**45**
liquid hammer *hammering* *pipe hammer* ○	**golpe de ariete**	coup de bélier	3.5.1-**46**
mains (supply) (UK) *power (supply) (USA)* ○	**alimentación eléctrica**	alimentation électrique	3.5.1-**47**
manual starting system	**sistema de arranque manual**	démarrage manuel	3.5.1-**48**
maximum allowable pressure *design pressure* *maximum declared pressure* *maximum working pressure (MWP)*	**presión máxima de servicio** presión de diseño presión máxima admisible presión máxima declarada presión máxima de trabajo	pression maximale de service	3.5.1-**49**
maximum declared pressure *design pressure* *maximum allowable pressure* *maximum working pressure (MWP)*	**presión máxima de servicio** presión de diseño presión máxima admisible presión máxima declarada presión máxima de trabajo	pression maximale de service	3.5.1-**50**

ENGLISH	ESPAÑOL	FRANÇAIS
3.5.1-**51**	maximum working pressure (MWP) *design pressure* *maximum allowable pressure* *maximum declared pressure*	**presión máxima de servicio** presión de diseño presión máxima admisible presión máxima declarada presión máxima de trabajo	pression maximale de service
3.5.1-**52**	monitoring	**supervisión** vigilancia	surveillance (technique)
3.5.1-**53**	multiple-speed governor controller	**regulador multiple de velocidad**	régulateur à vitesse multiple
3.5.1-**54**	needle hammer *hammering (of an expansion valve)* ○	**martilleo (de una válvula de expansión)** golpeteo (de une válvula de expansión)	martèlement d'un obturateur
3.5.1-**55**	no-load starting ○ *unloaded start* ○	**evacuación (de refrigerante)** arranque en vacío	démarrage à vide
3.5.1-**56**	non-condensable gas	**gas no condensable**	gaz non condensable *incondensable*
3.5.1-**57**	operating conditions	**régimen de funcionamiento**	régime de fonctionnement *régime de marche*
3.5.1-**58**	operating pressure *working pressure*	**presión de funcionamiento** presión de servicio	pression en service *pression de service*
3.5.1-**59**	operating temperature	**temperatura de funcionamiento**	température en service
3.5.1-**60**	overcharge	**exceso de fluido** exceso de carga	excès de charge *excès de fluide* ○
3.5.1-**61**	overfeeding	**sobrealimentación**	suralimentation
3.5.1-**62**	overflow	**rebosadero**	trop-plein
3.5.1-**63**	overheating	**calentamiento (anormal)**	échauffement (anormal) *surchauffe (d'un équipement)*
3.5.1-**64**	pipe hammer ○ *hammering* *liquid hammer*	**golpe de ariete**	coup de bélier
3.5.1-**65**	power (supply) (USA) ○ *mains (supply) (UK)*	**alimentación eléctrica**	alimentation électrique
3.5.1-**66**	power take-off	**toma de fuerza**	prise de force
3.5.1-**67**	pressure equalizing	**equilibrado de presiones**	équilibrage des pressions
3.5.1-**68**	pull-down test	**ensayo de puesta en régimen**	essai de mise en régime
3.5.1-**69**	pump down (of refrigerant)	**evacuación (de fluido frigorígeno)** vaciado	évacuation (du frigorigène)
3.5.1-**70**	purging *gas purging* *degassing* ○ *gas-off* ○	**purga de gas** purga	dégazage *purge*
3.5.1-**71**	rapid cycling	**funcionamiento en ciclos rápidos**	fonctionnement en cycles rapides
3.5.1-**72**	refrigerant charge	**carga de fluido frigorígeno**	charge de fluide (frigorigène)
3.5.1-**73**	refrigerant cylinder *gas bottle (1)* ○	**botella (de suministro) de fluido frigorífico** bottela de fluido frigorígeno botella de gas	bouteille de frigorigène (de livraison)
3.5.1-**74**	refrigeration running test	**ensayo frigorífico** ensayo de enfriamiento	essai frigorifique *essai de refroidissement*
3.5.1-**75**	running cycle	**ciclo de funcionamiento**	période
3.5.1-**76**	service charge ○ *holding charge*	**carga de servicio**	charge d'attente

ENGLISH	ESPAÑOL	FRANÇAIS	
service cylinder o *gas bottle (2)* o	**botella de fluido frigorífico (para mantenimiento)**	bouteille de frigorigène (de dépanneur)	3.5.1-**77**
short-cycling	**funcionamiento en ciclos cortos**	fonctionnement en courts cycles *courts cycles*	3.5.1-**78**
single-speed governor controller	**regulador de velocidad única**	régulateur à vitesse unique	3.5.1-**79**
slugging	**golpe de líquido**	coup de liquide	3.5.1-**80**
starting air valve	**valvula neumatica de arranque**	soupape de démarrage	3.5.1-**81**
starting interlock	**enclavamiento de arranque**	sécurité de démarrage	3.5.1-**82**
starting system	**sistema de arranque**	système de démarrage	3.5.1-**83**
starving (of an evaporator)	**subalimentación (de un evaporador)**	sous-alimentation (d'un évaporateur)	3.5.1-**84**
strength-test pressure	**presión de prueba** presión de ensayo	pression de l'essai de résistance *pression d'épreuve* o	3.5.1-**85**
test bed ○ *test bench* *test stand* *test rig* o	**banco de pruebas**	banc d'essai	3.5.1-**86**
test bench *test stand* *test rig* o *test bed* ○	**banco de pruebas**	banc d'essai	3.5.1-**87**
test rig o *test bench* *test stand* *test bed* ○	**banco de pruebas**	banc d'essai	3.5.1-**88**
test stand *test bench* *test rig* o *test bed* ○	**banco de pruebas**	banc d'essai	3.5.1-**89**
test temperature	**temperatura de prueba**	température lors de l'essai de résistance	3.5.1-**90**
trouble shooting	**diagnóstico**	diagnostic	3.5.1-**91**
undercharge *lack of refrigerant* *gas shortage* o	**falta de fluido**	manque de fluide	3.5.1-**92**
unloaded start o *no-load starting* o	**evacuación (de refrigerante)** arranque en vacío	démarrage à vide	3.5.1-**93**
unloader	**bypass de arranque** descargador	bipasse de démarrage *dispositif de délestage*	3.5.1-**94**
v-belt	**correa trapezoidal**	courroie trapézoïdale	3.5.1-**95**
vacuum test	**ensayo en vacío**	essai sous vide	3.5.1-**96**
valve bounce o *valve flutter*	**golpeteo (de una válvula)**	battement (d'un clapet)	3.5.1-**97**
valve flutter *valve bounce* o	**golpeteo (de una válvula)**	battement (d'un clapet)	3.5.1-**98**
vapour lock *gas lock*	**atasco por vapor** oclusión por vapor	bouchon de vapeur	3.5.1-**99**
variable-speed governor controller	**regulador de velocidad variable**	régulateur à vitesse variable *régulateur toute vitesse* o	3.5.1-**100**
working pressure *operating pressure*	**presión de funcionamiento** presión de servicio	pression en service *pression de service*	3.5.1-**101**

ENGLISH	ESPAÑOL	FRANÇAIS
SECTION 3.5.2 *Lubrication*	**SUBCAPÍTULO 3.5.2** *Lubricación*	**SOUS-CHAPITRE 3.5.2** *Lubrification*
3.5.2-1 autogenous ignition temperature ○ *spontaneous ignition temperature*	**temperatura de autoinflamación** temperatura de inflamación espontánea	température d'auto-inflammation
3.5.2-2 breakdown (of an oil)	**descomposición (de un aceite)**	décomposition (d'une huile) *altération*
3.5.2-3 carbonization	**carbonización**	carbonisation
3.5.2-4 cloud point	**punto de enturbiamiento**	point de trouble
3.5.2-5 dip lubrication	**lubricación por barboteo**	lubrification par barbotage
3.5.2-6 drop point	**punto de gota**	point de goutte
3.5.2-7 fire point	**punto de combustión**	point d'inflammation
3.5.2-8 flash point	**punto de inflamación**	point d'éclair
3.5.2-9 flock point	**punto de floculación**	point de floculation
3.5.2-10 foaming	**formación de espuma**	moussage
3.5.2-11 forced-feed oiling ○ *forced lubrication* *mechanical lubrication ◑* *pump lubrication ◑*	**engrase a presión** lubricación a presión lubricación forzada	lubrification sous pression *lubrification forcée*
3.5.2-12 forced lubrication *mechanical lubrication ◑* *pump lubrication ◑* *forced-feed oiling ○*	**engrase a presión** lubricación a presión lubricación forzada	lubrification sous pression *lubrification forcée*
3.5.2-13 lubrication	**engrase** lubricación	lubrification *graissage*
3.5.2-14 mechanical lubrication ◑ *forced lubrication* *pump lubrication ◑* *forced-feed oiling ○*	**engrase a presión** lubricación a presión lubricación forzada	lubrification sous pression *lubrification forcée*
3.5.2-15 oil charge	**carga de aceite**	charge d'huile
3.5.2-16 oil content	**contenido de aceite**	teneur en huile
3.5.2-17 oil cooler	**enfriador de aceite**	refroidisseur d'huile
3.5.2-18 oil distributor	**circuito de distribución de aceite**	circuit d'huile (d'une machine)
3.5.2-19 oil gauge ○ *oil sight glass*	**indicador de nivel de aceite**	indicateur de niveau d'huile *niveau d'huile ◑*
3.5.2-20 oil level	**nivel de aceite**	niveau d'huile
3.5.2-21 oil level indicator	**visor de nivel de aceite**	indicateur de niveau d'huile *niveau d'huile ◑*
3.5.2-22 oil pressure	**presión de aceite**	pression d'huile
3.5.2-23 oil pressure gauge	**manómetro de presión de aceite**	manomètre de pression d'huile
3.5.2-24 oil pump	**bomba de aceite**	pompe à huile
3.5.2-25 oil removal	**extracción de aceite**	déshuilage
3.5.2-26 oil return	**retorno de aceite**	retour d'huile
3.5.2-27 oil sight glass *oil gauge ○*	**indicador de nivel de aceite**	indicateur de niveau d'huile *niveau d'huile ◑*

ENGLISH	ESPAÑOL	FRANÇAIS	
pour point	punto de goteo	point d'écoulement	3.5.2-28
pump lubrication ◐ *forced lubrication* mechanical lubrication ◐ *forced-feed oiling* ○	engrase a presión lubricación a presión lubricación forzada	lubrification sous pression *lubrification forcée*	3.5.2-29
sludge	fangos	boues	3.5.2-30
softening point	punto de reblandecimiento	point de ramollissement	3.5.2-31
splash lubrication	engrase por barboteo engrase por salpicadura	lubrification par projection	3.5.2-32
spontaneous ignition temperature *autogenous ignition temperature* ○	temperatura de autoinflamación temperatura de inflamación espontánea	température d'auto-inflammation	3.5.2-33
viscosity index	índice de viscosidad	indice de viscosité	3.5.2-34
wax content	contenido de parafina	teneur en paraffine	3.5.2-35

SECTION 3.5.3 *Defrosting*	SUBCAPÍTULO 3.5.3 *Desescarche*	SOUS-CHAPITRE 3.5.3 *Dégivrage*	
automatic defrosting	desescarche automático	dégivrage automatique	3.5.3-1
defrost pan (USA) *drip tray*	bandeja de desecarche cubeta de desescarche	bac de dégivrage *cuvette de dégivrage*	3.5.3-2
defrost time	tiempo de descarche	durée de dégivrage	3.5.3-3
defrost water removal	eliminación del agua de desescarche	évacuation de l'eau de dégivrage	3.5.3-4
defrosting	desescarche	dégivrage	3.5.3-5
defrosting cycle	ciclo de desescarche	cycle de dégivrage	3.5.3-6
drip tray *defrost pan (USA)*	bandeja de desecarche cubeta de desescarche	bac de dégivrage *cuvette de dégivrage*	3.5.3-7
electric defrosting	desescarche eléctrico	dégivrage électrique	3.5.3-8
external defrosting	desescarche desde el exterior	dégivrage par l'extérieur	3.5.3-9
frost	escarcha	givre	3.5.3-10
frost back ◐	formación de escarcha (en la aspiración)	givrage à l'aspiration (d'un compresseur) ◐	3.5.3-11
frost deposit ○	depósito de escarcha	dépôt de givre ○	3.5.3-12
frost formation	formación de escarcha	givrage	3.5.3-13
frosted ◐	escarchado (adj.)	givré ◐	3.5.3-14
hot-gas defrosting	desescarche por gases calientes	dégivrage par gaz chauds	3.5.3-15
internal defrosting	desescarche desde el interior	dégivrage par l'intérieur	3.5.3-16
manual defrosting	desescarche manual	dégivrage manuel	3.5.3-17
off-cycle defrosting	desescarche natural cíclico desescarche natural periódico	dégivrage naturel cyclique *dégivrage à chaque cycle*	3.5.3-18
pressure defrosting	desescarche gobernado por la caída de presión en el aire	dégivrage commandé par la perte de pression sur l'air	3.5.3-19
reverse-cycle defrosting	desescarche por inversión de ciclo	dégivrage par cycle inversé *dégivrage par inversion de cycle*	3.5.3-20
semi-automatic defrosting	desescarche semiautomático	dégivrage semi-automatique	3.5.3-21

	ENGLISH	ESPAÑOL	FRANÇAIS
3.5.3-**22**	thermobank defrost	**termoacumulador para desescarche**	thermo-accumulateur pour dégivrage
3.5.3-**23**	time defrosting	**desescarche temporizado**	dégivrage chronocommandé *dégivrage (commandé) par chronorelais*
3.5.3-**24**	time-temperature defrosting	**desescarche gobernado por temperatura-tiempo**	dégivrage à commande combinée par le temps et la température *dégivrage chrono-thermique*
3.5.3-**25**	water defrosting	**desescarche por (asperción de) agua** desescarche por (lluvia de) agua	dégivrage par (aspersion d') eau *dégivrage par (ruissellement d') eau*

Capítulo 4.

MÉTODOS DE ENFRIAMIENTO

● término aceptado

○ término obsoleto

ENGLISH	ESPAÑOL	FRANÇAIS	
SECTION 4.1 *Cooling methods:* *general background*	**SUBCAPÍTULO 4.1** *Generalidades sobre métodos* *de enfriamiento*	**SOUS-CHAPITRE 4.1** *Généralités sur les méthodes* *de refroidissement*	
air blast cooling	**enfriamiento por corriente de aire** enfriamiento por flujo de aire	refroidissement par air soufflé *refroidissement par soufflage d'air*	4.1-**1**
contact cooling	**enfriamiento por contacto**	refroidissement par contact	4.1-**2**
cooling bath ○	**baño de enfriamiento**	bain de refroidissement	4.1-**3**
cooling method *cooling process* ◐	**método de enfriamiento**	méthode de refroidissement *mode de refroidissement*	4.1-**4**
cooling process ◐ *cooling method*	**método de enfriamiento**	méthode de refroidissement *mode de refroidissement*	4.1-**5**
direct expansion refrigeration *direct refrigerating system*	**sistema de enfriamiento directo** enfriamiento por expansión directa sistema frigorifero directo	système de refroidissement direct *refroidissement par détente directe*	4.1-**6**
direct refrigerating system *direct expansion refrigeration*	**sistema de enfriamiento directo** enfriamiento por expansión directa sistema frigorifero directo	système de refroidissement direct *refroidissement par détente directe*	4.1-**7**
equilibration temperature	**temperatura de equilibrio**	température d'équilibre *température d'équilibre (thermique)*	4.1-**8**
evaporative cooling	**enfriamiento evaporativo** enfriamiento por evaporación de agua	refroidissement évaporatif *refroidissement par évaporation d'eau*	4.1-**9**
forced-draught cooling ○	**enfriamiento por convección forzada** enfriamiento por circulación forzada	refroidissement par convection forcée *refroidissement par ventilation forcée*	4.1-**10**
immersion cooling	**enfriamiento por inmersión**	refroidissement par immersion	4.1-**11**
indirect refrigerating system	**sistema de enfriamiento indirecto** sistema de enfriamiento por transmisión indirecta sistema de enfriamiento secundario sistema frigorifero indirecto	système de refroidissement indirect *système secondaire de refroidissement* *système à fluide secondaire* ◐	4.1-**12**
natural-convection cooling	**enfriamiento por convección natural**	refroidissement par convection naturelle	4.1-**13**
partial-recovery refrigeration	**enfriamiento con fluido frigorígeno parcialmente recuperado**	refroidissement à frigorigène partiellement récupéré	4.1-**14**
recessed fitting	**montaje empotrado**	montage encastré	4.1-**15**
spray cooling	**enfriamiento por aspersión** enfriamiento por lluvia	refroidissement par aspersion	4.1-**16**
surface cooling	**enfriamiento mediante superficie fría**	refroidissement sur surface froide	4.1-**17**
total-loss refrigeration	**enfriamiento con fluido frigorígeno perdido**	refroidissement à frigorigène perdu	4.1-**18**
transpiration cooling	**enfriamiento por transpiración**	refroidissement par transpiration	4.1-**19**
vacuum chilling ◐ *vacuum cooling*	**enfriamiento por vacío**	réfrigération par le vide *refroidissement par le vide*	4.1-**20**
vacuum cooling *vacuum chilling* ◐	**enfriamiento por vacío**	réfrigération par le vide *refroidissement par le vide*	4.1-**21**

ENGLISH	ESPAÑOL	FRANÇAIS	
SECTION 4.2 *Chilling*	**SUBCAPÍTULO 4.2** *Refrigeración*	**SOUS-CHAPITRE 4.2** *Réfrigération*	
chill (to)	**refrigerar**	réfrigérer	4.2-**1**
chilling	**refrigeración**	réfrigération	4.2-**2**

	ENGLISH	ESPAÑOL	FRANÇAIS
4.2-3	cooling down	**puesta en temperatura**	opération de refroidissement *tombée en froid* ○
4.2-4	cooling rate	**velocidad de enfriamiento**	vitesse de refroidissement
4.2-5	flood-type hydrocooling	**refrigeración por inmersión en agua helada**	réfrigération par immersion dans l'eau glacée
4.2-6	fogging	**brumización**	brumisation *nébulisation* ○
4.2-7	half-cooling time	**tiempo de semienfriamiento**	temps de demi-refroidissement
4.2-8	hydrocooling	**refrigeración por agua helada**	réfrigération par eau glacée
4.2-9	impingement cooling *jet cooling* ○	**enfriamiento por chorro de aire frío**	refroidissement par jet d'air froid
4.2-10	jacket cooling	**refrigeración por doble pared**	réfrigération par double paroi *réfrigération par enveloppe froide* ○
4.2-11	jet cooling ○ *impingement cooling*	**enfriamiento por chorro de aire frío**	refroidissement par jet d'air froid
4.2-12	liquid-chilling package	**unidad enfriadora de líquidos**	groupe refroidisseur de liquide
4.2-13	precooler	**preenfriador**	prérefroidisseur
4.2-14	precooling	**prerrefrigeración**	préréfrigération
4.2-15	pressure cooling	**refrigeración por presión de aire**	réfrigération par pression d'air
4.2-16	quick chilling ○ *rapid chilling*	**refrigeración rápida**	réfrigération rapide
4.2-17	rapid chilling *quick chilling* ○	**refrigeración rápida**	réfrigération rapide
4.2-18	refrigerator	**refrigerador**	réfrigérateur
4.2-19	spray-type hydrocooling	**refrigeración por aspersión de agua helada** refrigeración por lluvia de agua helada	réfrigération par aspersion d'eau glacée
4.2-20	superchilling	**refrigeración crítica** refrigeración límite	superréfrigération

	SECTION 4.3 *Freezing*	SUBCAPÍTULO 4.3 *Congelación*	SOUS-CHAPITRE 4.3 *Congélation*
4.3-1	air-blast freezer	**congelador por aire frío**	congélateur à air soufflé
4.3-2	air-blast freezing	**congelación por aire frío**	congélation par air soufflé
4.3-3	batch freezer	**congelador de una capa** congelador monocapa discontínuo	congélateur discontinu
4.3-4	belt freezer	**congelador de cinta**	congélateur à convoyeur
4.3-5	belt freezing	**congelación en contínuo por aire frío**	congélation sur convoyeur
4.3-6	bulk freezing	**congelación a granel**	congélation en vrac
4.3-7	contact freezer	**congelador por contacto**	congélateur par contact
4.3-8	contact freezing	**congelación por contacto**	congélation par contact
4.3-9	continuous freezer	**congelador contínuo**	congélateur continu
4.3-10	crust freezing *shell freezing* ○	**congelación superficial**	croûtage *congélation superficielle* ○
4.3-11	cryogenic freezer	**congelador criogénico**	congélateur (à froid) cryogénique

ENGLISH	ESPAÑOL	FRANÇAIS	
deep freezing ○ *quick freezing*	**congelación rápida**	surgélation *congélation rapide ○*	4.3-**12**
deep-freezing plant ○ *quick-freezing plant*	**planta de congelación rápida**	installation de surgélation *installation de congélation rapide ○*	4.3-**13**
deep-frozen food ○ *quick-frozen food*	**alimento ultracongelado**	denrées surgelées *aliments surgelés*	4.3-**14**
dehydro-freezing	**desecado y congelación** desidratación y congelación	déshydratation-congélation	4.3-**15**
dielectric thawing	**descongelación dieléctrica**	décongélation diélectrique	4.3-**16**
double-contact freezer	**congelador por doble contacto**	congélateur par double contact	4.3-**17**
drip	**exudado**	exsudat	4.3-**18**
dual freezing system	**congelador dual** congelador por dos métodos	congélateur à froid mixte	4.3-**19**
effective freezing time	**duración efectiva de congelación**	durée effective de congélation *temps effectif de congélation*	4.3-**20**
fluidized bed	**lecho fluidizado**	lit fluidisé	4.3-**21**
fluidized-bed freezer	**congelador en lecho fluidizado**	congélateur à lit fluidisé	4.3-**22**
fluidized-bed freezing *fluidized freezing ○*	**congelación en lecho fluidizado**	congélation en lit fluidisé	4.3-**23**
fluidized freezing ○ *fluidized-bed freezing*	**congelación en lecho fluidizado**	congélation en lit fluidisé	4.3-**24**
freeze (to) (foods)	**congelar**	congeler (des aliments)	4.3-**25**
freezer *freezer unit ○*	**congelador**	congélateur	4.3-**26**
freezer capacity	**capacidad del congelador**	charge d'un congélateur	4.3-**27**
freezer unit ○ *freezer*	**congelador**	congélateur	4.3-**28**
freezing capacity	**capacidad de congelación**	capacité de congélation	4.3-**29**
freezing method *freezing process ○*	**método de congelación**	méthode de congélation *mode de congélation*	4.3-**30**
freezing plant	**planta de congelación**	installation de congélation	4.3-**31**
freezing plateau	**meseta de congelación (en un diagrama de temperatura-tiempo)** palier de congelación (en un diagrama de temperatura-tiempo)	palier de congélation	4.3-**32**
freezing process ○ *freezing method*	**método de congelación**	méthode de congélation *mode de congélation*	4.3-**33**
freezing rate *speed of freezing ○*	**velocidad de congelación**	vitesse de congélation	4.3-**34**
freezing room	**cámara de congelación**	chambre de congélation	4.3-**35**
freezing section	**sección de congelación**	atelier de congélation	4.3-**36**
freezing time	**tiempo de congelación**	durée de congélation	4.3-**37**
freezing tunnel *tunnel freezer ○*	**túnel de congelación**	tunnel de congélation	4.3-**38**
frozen food	**alimento congelado**	aliment congelé *denrée congelée*	4.3-**39**
frozen-food locker ○	**compartimiento individual para artículos congelados**	casier congélateur *casier pour denrées congelées*	4.3-**40**

	ENGLISH	ESPAÑOL	FRANÇAIS
4.3-41	glazing	**glaseado**	glaçage *givrage*
4.3-42	high-frequency thawing	**descongelación dieléctrica por alta frecuencia**	décongélation par haute fréquence
4.3-43	ice-slurry freezing	**congelación en hielo líquido**	congélation dans un coulis de glace
4.3-44	immersion freezer	**congelador por inmersión**	congélateur par immersion
4.3-45	immersion freezing	**congelación por inmersión**	congélation par immersion
4.3-46	impingement freezing *jet freezing* ○	**congelación por chorro de aire**	congélation par jet d'air
4.3-47	individual quick freezing (IQF)	**congelación individual ultrarrápida**	surgélation individuelle
4.3-48	initial freezing temperature ● *temperature of initial ice formation*	**temperatura inicial de congelación** temperatura de principio de congelación	température de congélation commençante
4.3-49	initial melting temperature	**temperatura de inicio de la descongelación**	température de fusion commençante
4.3-50	jet freezing ○ *impingement freezing*	**congelación por chorro de aire**	congélation par jet d'air
4.3-51	lift off (of particles)	**pérdida (de particulas)**	envol (de particules)
4.3-52	loose freezing	**congelación de productos sueltos**	congélation en masse divisée
4.3-53	mechanical freezer	**congelador por frío mecánico**	congélateur à froid mécanique
4.3-54	microwave thawing *ultra-high frequency thawing* ○	**descongelación por microondas**	décongélation par microondes *décongélation par hyperfréquence* ●
4.3-55	multiplate freezer	**congelador de placas múltiples** congelador multiplaca	congélateur à plaques multiples
4.3-56	nominal freezing time	**tiempo nominal de congelación**	durée nominale de congélation
4.3-57	plate freezer	**congelador de placas**	congélateur à plaque
4.3-58	quick freezing *deep freezing* ●	**congelación rápida**	surgélation *congélation rapide* ●
4.3-59	quick-freezing plant *deep-freezing plant* ●	**planta de congelación rápida**	installation de surgélation *installation de congélation rapide* ●
4.3-60	quick-frozen food *deep-frozen food* ●	**alimento ultracongelado**	denrées surgelées *aliments surgelés*
4.3-61	rated freezing capacity	**capacidad nominal de congelación**	aptitude de congélation nominale
4.3-62	refreezing	**recongelación**	recongélation
4.3-63	refrigerated locker ○	**compartimiento frigorífico individual**	case frigorifique *casier frigorifique*
4.3-64	scraped-surface freezer	**congelador con rascador**	congélateur à surface raclée *congélateur à racleur* ●
4.3-65	shelf freezer	**congelador de estanterías**	congélateur à étagères
4.3-66	shell freezing ● *crust freezing*	**congelación superficial**	croûtage *congélation superficielle* ●
4.3-67	slow freezing	**congelación lenta**	congélation lente
4.3-68	speed of freezing ● *freezing rate*	**velocidad de congelación**	vitesse de congélation
4.3-69	spray freezer	**congelador por aspersión**	congélateur par aspersion *congélateur par pulvérisation*
4.3-70	spray freezing	**congelación por pulverización** congelación por aspersión congelación por atomización	congélation par aspersion *congélation par pulvérisation*

ENGLISH	ESPAÑOL	FRANÇAIS	
still-air freezing	**congelación en aire en reposo**	congélation en air calme	4.3-**71**
temperature of initial ice formation *initial freezing temperature* ○	**temperatura inicial de congelación** temperatura de principio de congelación	température de congélation commençante	4.3-**72**
thaw (to)	**descongelar**	décongeler *dégeler*	4.3-**73**
thawing	**descongelación**	décongélation	4.3-**74**
thawing time	**tiempo de descongelación**	durée de décongélation	4.3-**75**
thermal centre	**centro térmico**	centre thermique	4.3-**76**
tunnel freezer ○ *freezing tunnel*	**túnel de congelación**	tunnel de congélation	4.3-**77**
ultra-high frequency thawing ○ *microwave thawing*	**descongelación por microondas**	décongélation par microondes *décongélation par hyperfréquence* ○	4.3-**78**

SECTION 4.4 *Vacuum application*	SUBCAPÍTULO 4.4 *Tratamiento al vacío*	SOUS-CHAPITRE 4.4 *Traitement sous vide*	
cold trap	**condensador de trampa de frío**	piège froid	4.4-**1**
condensation (vacuum) pump ○ *diffusion (vacuum) pump*	**bomba (de vacío) de difusión** bomba (de vacío) de condensación	pompe (à vide) à condensation *pompe (à vide) à diffusion* ○	4.4-**2**
diffusion (vacuum) pump *condensation (vacuum) pump* ○	**bomba (de vacío) de difusión** bomba (de vacío) de condensación	pompe (à vide) à condensation *pompe (à vide) à diffusion* ○	4.4-**3**
flash point	**punto de evaporación** punto de vaporización	point de vaporisation *point d'évaporation*	4.4-**4**
ion (vacuum) pump	**bomba (de vacío) iónica**	pompe (à vide) ionique	4.4-**5**
liquid mechanical (vacuum) pump	**bomba (de vacío) de anillo líquido**	pompe (à vide) à anneau liquide	4.4-**6**
sorption (vacuum) pump	**bomba (de vacío) de sorción**	pompe (à vide) à sorption	4.4-**7**
vacuum breaking *vacuum cracking* ○	**ruptura del vacío**	cassage du vide	4.4-**8**
vacuum cracking ○ *vacuum breaking*	**ruptura del vacío**	cassage du vide	4.4-**9**
vacuum device	**dispositivo de vacío**	appareil sous vide	4.4-**10**
vacuum plant ○ *vacuum system*	**instalación de vacío**	installation de vide	4.4-**11**
vacuum pump	**bomba de vacío**	pompe à vide	4.4-**12**
vacuum system *vacuum plant* ○	**instalación de vacío**	installation de vide	4.4-**13**
vapour (vacuum) pump	**bomba (de vacío) de chorro de vapor**	pompe (à vide) à jet de vapeur *pompe à vapeur de... (mercure, huile...)*	4.4-**14**

Capítulo 5. | Almacenamiento, transporte, distribución

● término aceptado

○ término obsoleto

ENGLISH	ESPAÑOL	FRANÇAIS	
SECTION 5.1 *Domestic equipment*	**SUBCAPÍTULO 5.1** *Equipos domésticos*	**SOUS-CHAPITRE 5.1** *Equipements ménagers*	
absorption refrigerator	**refrigerador de absorción** frigorífico de absorción	réfrigérateur à absorption	5.1-**1**
beverage dispenser	**distribuidor de bebidas**	distributeur de boissons	5.1-**2**
bottle beverage cooler *bottle cooler* ○	**enfriador de botellas**	refroidisseur de bouteilles	5.1-**3**
bottle cooler ○ *bottle beverage cooler*	**enfriador de botellas**	refroidisseur de bouteilles	5.1-**4**
built-in refrigerator	**refrigerador empotrado** frigorífico empotrado	réfrigérateur encastré *réfrigérateur intégré* *réfrigérateur incorporé* ○	5.1-**5**
cooling device	**dispositivo de enfriamiento**	dispositif de refroidissement	5.1-**6**
domestic freezer *household food freezer* *home freezer* ○	**congelador doméstico** congelador	congélateur ménager *congélateur domestique* ○	5.1-**7**
domestic refrigerator *household refrigerator*	**refrigerador doméstico** frigorífico doméstico	réfrigérateur ménager *armoire frigorifique ménagère* *réfrigérateur domestique* ○	5.1-**8**
door racks	**estanterías de la contrapuerta**	balconnet	5.1-**9**
drinking-water cooler *fountain (USA)* ○ *water dispenser* ○	**enfriador de agua potable**	rafraîchisseur d'eau (potable) *fontaine réfrigérée*	5.1-**10**
dual-temperature refrigerator	**refrigerador de dos temperaturas**	réfrigérateur-congélateur	5.1-**11**
enamel (paint)	**pintura al esmalte**	peinture laquée	5.1-**12**
fountain (USA) ○ *drinking water cooler* *water dispenser* ○	**enfriador de agua potable**	rafraîchisseur d'eau (potable) *fontaine réfrigérée*	5.1-**13**
freezer compartment *low-temperature compartment* ○	**compartimento congelador**	compartiment congélateur *compartiment à basse température*	5.1-**14**
freezer total gross volume	**volumen bruto total**	volume brut total d'un congélateur	5.1-**15**
freezer total storage volume	**volumen total de almacenamiento** volumen total útil	volume utile total d'un congélateur	5.1-**16**
gas refrigerator	**refrigerador por gas** frigorífico por gas	réfrigérateur à gaz	5.1-**17**
home freezer ○ *domestic freezer* *household food freezer*	**congelador doméstico** congelador	congélateur ménager *congélateur domestique* ○	5.1-**18**
household food freezer *domestic freezer* *home freezer* ○	**congelador doméstico** congelador	congélateur ménager *congélateur domestique* ○	5.1-**19**
household frozen-food storage cabinet ("three-star" cabinet)	**congelador**	conservateur conservateur de denrées congelées à usage ménager (conservateur "trois étoiles")	5.1-**20**
household refrigerator *domestic refrigerator*	**refrigerador doméstico** frigorífico doméstico	réfrigérateur ménager *armoire frigorifique ménagère* *réfrigérateur domestique* ○	5.1-**21**
ice box ○ *ice refrigerator*	**nevera doméstica** refrigerador de hielo	glacière domestique *réfrigérateur à glace*	5.1-**22**
ice (cube) tray	**bandeja para hielo**	tiroir à glace *tiroir à glaçons*	5.1-**23**

	ENGLISH	ESPAÑOL	FRANÇAIS
5.1-24	ice refrigerator *ice box* ○	**nevera doméstica** refrigerador de hielo	glacière domestique *réfrigérateur à glace*
5.1-25	inner cabinet liner *liner (of a refrigerator)* ○	**cuba interior (de un refrigerador)**	cuve intérieure (d'un réfrigérateur) *cuve interne (d'un réfrigérateur)* ○
5.1-26	liner (of a refrigerator) ○ *inner cabinet liner*	**cuba interior (de un refrigerador)**	cuve intérieure (d'un réfrigérateur) *cuve interne (d'un réfrigérateur)* ○
5.1-27	load-limit volume ○	**volumen interior bruto** volumen límite de carga (capacidad)	volume intérieur brut *volume de charge maximal*
5.1-28	low-temperature compartment ○ *freezer compartment*	**compartimento congelador**	compartiment congélateur *compartiment à basse température*
5.1-29	nursery refrigerator	**enfriador de biberones**	réfrigérateur de biberons
5.1-30	open refrigerator	**expositor frigorífico**	présentoir frigorifique
5.1-31	overall dimensions (doors or lids closed)	**dimensiones totales**	dimensions hors-tout (portes ou couvercles fermés) *hors-tout*
5.1-32	overall space required in use (doors or lids open)	**dimensiones en uso**	encombrement en service (portes ou couvercles ouverts)
5.1-33	rated energy consumption	**consumo energético**	consommation d'énergie nominale
5.1-34	rated gross volume	**volumen bruto nominal**	volume brut nominal
5.1-35	rated storage volume	**volumen útil nominal**	volume utile nominal
5.1-36	rated total gross volume	**volumen bruto total nominal**	volume brut total nominal
5.1-37	rated total storage volume	**volumen útil total nominal**	volume utile total nominal
5.1-38	self-contained refrigerator	**enfriador con grupo incorporado**	réfrigérateur à groupe incorporé
5.1-39	soda-fountain ○	**fuenta de soda**	fontaine à soda
5.1-40	storage shelf area	**superficie útil de almacenamiento**	surface utile de rangement
5.1-41	three-star section	**sección tres estrellas**	partie "trois étoiles"
5.1-42	top-opening type	**cofre congelador**	appareil du type coffre
5.1-43	two-star section	**sección dos estrellas**	partie "deux étoiles"
5.1-44	upright freezer	**congelador doméstico tipo armario**	congélateur vertical *congélateur-armoire*
5.1-45	upright type	**armario congelador**	appareil du type armoire
5.1-46	vitreous enamel	**esmalte vitrificado**	émail vitrifié
5.1-47	water dispenser ○ *drinking water cooler* *fountain (USA)* ○	**enfriador de agua potable**	rafraîchisseur d'eau (potable) *fontaine réfrigérée*

	SECTION 5.2 *Commercial equipment*	SUBCAPÍTULO 5.2 *Equipos comerciales*	SOUS-CHAPITRE 5.2 *Equipements commerciaux*
5.2-1	air curtain	**cortina de aire**	rideau d'air
5.2-2	air outlet	**salida de aire**	soufflage d'air
5.2-3	air return	**toma de aire de retorno**	reprise d'air
5.2-4	assisted-service cabinet	**mueble mostrador**	meuble à service assisté
5.2-5	automatic food-vending machine	**distribuidor automático de productos alimenticios**	distributeur automatique de denrées

ENGLISH	ESPAÑOL	FRANÇAIS	
base	**bandeja inferior del expositor**	plan d'exposition inférieur	5.2-**6**
cabinet shell	**cuba exterior (de un mueble frigorífico)**	enveloppe (d'un meuble frigorifique) *cuve extérieure (d'un meuble frigorifique)* ○	5.2-**7**
canopy	**frontal superior**	fronton	5.2-**8**
closed refrigerated display cabinet	**expositor refrigerado con puerta**	meuble frigorifique de présentation fermé	5.2-**9**
commercial cabinet *commercial refrigerator* ○ *reach-in refrigerator* ○ *service cabinet* ○	**armario frigorífico (comercial)**	armoire frigorifique commerciale *réfrigérateur commercial*	5.2-**10**
commercial refrigerator ○ *commercial cabinet* *reach-in refrigerator* ○ *service cabinet* ○	**armario frigorífico (comercial)**	armoire frigorifique commerciale *réfrigérateur commercial*	5.2-**11**
counter ○ *display cabinet*	**mueble expositor** mueble exposición vitrina	meuble de vente *comptoir* *meuble d'étalage* *meuble d'exposition* *présentoir* *vitrine* ○	5.2-**12**
display *display opening*	**expositor**	exposition *étalage* *zone d'exposition* *zone de présentation*	5.2-**13**
display cabinet *counter* ○	**mueble expositor** mueble exposición vitrina	meuble de vente *comptoir* *meuble d'étalage* *meuble d'exposition* *présentoir* *vitrine* ○	5.2-**14**
display opening *display*	**expositor**	exposition *étalage* *zone d'exposition* *zone de présentation*	5.2-**15**
display opening area	**superficie de exposición**	surface de la zone d'exposition	5.2-**16**
end cabinet	**cabecera de góndola**	meuble tête de gondole	5.2-**17**
end wall	**pared de cierre**	joue *panneau d'extrémité*	5.2-**18**
external angle	**góndola esquina**	meuble d'angle ouvert	5.2-**19**
front	**frente**	face avant	5.2-**20**
front riser	**tope de producto**	arrêt produit *arrêtoir*	5.2-**21**
frozen-food cabinet	**conservador de productos congelados**	conservateur pour produits congelés	5.2-**22**
frozen-food display case ○	**vitrina para productos congelados**	comptoir pour produits congelés	5.2-**23**
handrail	**protector de aristas**	main courante	5.2-**24**
horizontal refrigerated display cabinet	**góndola refrigerada horizontal**	meuble frigorifique de vente horizontal (MFV horizontal)	5.2-**25**
internal angle	**góndola esquina**	meuble d'angle fermé	5.2-**26**
internal gross volume	**volumen interior bruto**	volume intérieur brut	5.2-**27**
internal net volume	**volumen interior neto**	volume intérieur net	5.2-**28**
island cabinet *island display case* ○	**góndola**	gondole *meuble îlot*	5.2-**29**

	ENGLISH	ESPAÑOL	FRANÇAIS
5.2-**30**	island display case ◐ *island cabinet*	**góndola**	gondole *meuble îlot*
5.2-**31**	island run	**isla**	îlot
5.2-**32**	kickplate	**rodapie**	plinthe
5.2-**33**	line-up	**lineal de expositores**	linéaire
5.2-**34**	load limit	**límite de carga**	limite de chargement
5.2-**35**	load limit line	**línea límite de carga**	ligne de limite de chargement
5.2-**36**	night cover	**tapa de noche (de una vitrina)**	protecteur de nuit *couvercle de nuit*
5.2-**37**	normal conditions of use	**condiciones normales de uso**	conditions normales d'emploi
5.2-**38**	open refrigerated display cabinet	**góndola refrigerada abierta**	meuble frigorifique ouvert
5.2-**39**	overall external dimensions at installation	**espacio requerido en la instalación**	encombrement hors-tout à l'installation
5.2-**40**	overall external dimensions in service	**espacio requerido en la instalación en servicio**	encombrement hors-tout en service
5.2-**41**	price-marking rail	**porta etiqueta**	porte-étiquettes
5.2-**42**	reach-in refrigerator ◐ *commercial cabinet* *commercial refrigerator* ◐ *service cabinet* ◐	**armario frigorífico (comercial)**	armoire frigorifique commerciale *réfrigérateur commercial*
5.2-**43**	refrigerated cabinet ○	**armario frigorífico** mueble frigorífico	armoire frigorifique *meuble frigorifique*
5.2-**44**	refrigerated counter ◐ *refrigerated display cabinet* *refrigerated display case* ◐ *refrigerated showcase* ◐	**vitrina frigorífica**	meuble frigorifique de vente (MFV) *vitrine frigorifique* *présentoir frigorifique* ◐ *comptoir frigorifique* ○
5.2-**45**	refrigerated display cabinet *refrigerated counter* ◐ *refrigerated display case* ◐ *refrigerated showcase* ◐	**vitrina frigorífica**	meuble frigorifique de vente (MFV) *vitrine frigorifique* *présentoir frigorifique* ◐ *comptoir frigorifique* ○
5.2-**46**	refrigerated display case ◐ *refrigerated display cabinet* *refrigerated counter* ◐ *refrigerated showcase* ◐	**vitrina frigorífica**	meuble frigorifique de vente (MFV) *vitrine frigorifique* *présentoir frigorifique* ◐ *comptoir frigorifique* ○
5.2-**47**	refrigerated shelf area	**superficie de estanterías de un mueble**	surface des étagères d'un meuble frigorifique
5.2-**48**	refrigerated showcase ◐ *refrigerated display cabinet* *refrigerated counter* ◐ *refrigerated display case* ◐	**vitrina frigorífica**	meuble frigorifique de vente (MFV) *vitrine frigorifique* *présentoir frigorifique* ◐ *comptoir frigorifique* ○
5.2-**49**	refrigerated window	**escaparate frigorífico**	devanture frigorifique *vitrine frigorifique*
5.2-**50**	retro-case	**mueble con servicio posterior**	meuble à service arrière
5.2-**51**	roll-in cabinet	**mueble con frente accesible**	meuble à façade levable
5.2-**52**	roll-in refrigerator	**recinto frigorífico para carretillas**	armoire frigorifique à chariots
5.2-**53**	self-service cabinet *self-service display case* ◐	**góndola de autoservicio** vitrina de autoservicio	meuble libre-service *comptoir libre-service*
5.2-**54**	self-service display case ◐ *self-service cabinet*	**góndola de autoservicio** vitrina de autoservicio	meuble libre-service *comptoir libre-service*
5.2-**55**	semi-vertical refrigerated display cabinet	**mueble frigorífico de media altura**	meuble frigorifique de vente semi-vertical

ENGLISH	ESPAÑOL	FRANÇAIS	
service cabinet ● *commercial cabinet* *commercial refrigerator* ● *reach-in refrigerator* ●	**armario frigorífico (comercial)**	armoire frigorifique commerciale *réfrigérateur commercial*	5.2-**56**
shelf	**estante**	étagère	5.2-**57**
shelf sham	**falso volumen**	fausse masse	5.2-**58**
storage volume	**volumen de almacenamiento**	volume utile	5.2-**59**
technical line-up	**lineal multitemperatura de expositores**	linéaire technique	5.2-**60**
vertical refrigerated display cabinet	**mueble frigorífico vertical con estantes**	meuble frigorifique de vente vertical	5.2-**61**

SECTION 5.3 *Cold rooms*	SUBCAPÍTULO 5.3 *Cámaras frigoríficas*	SOUS-CHAPITRE 5.3 *Chambres froides*	
air lock	**antecámara de esclusa** esclusa de aire	sas d'air	5.3-**1**
anteroom	**antecámara**	antichambre	5.3-**2**
capacity of a cold room	**capacidad de una cámara frigorífica**	capacité d'une chambre froide	5.3-**3**
chill room ● *chilling room* *cooler (USA)* ●	**cámara de refrigeración**	chambre de réfrigération	5.3-**4**
chilling room *chill room* ● *cooler (USA)* ●	**cámara de refrigeración**	chambre de réfrigération	5.3-**5**
cold chamber ● *cold room*	**cámara frigorífica**	chambre froide	5.3-**6**
cold room *cold chamber* ●	**cámara frigorífica**	chambre froide	5.3-**7**
cold storage room	**cámara de almacenamiento frigorífico**	chambre d'entreposage frigorifique *chambre de stockage frigorifique* ●	5.3-**8**
cooler (USA) ● *chilling room* *chill room* ●	**cámara de refrigeración**	chambre de réfrigération	5.3-**9**
cooling tunnel	**túnel de refrigeración**	tunnel de réfrigération	5.3-**10**
effective capacity ● *net capacity*	**volumen útil de una cámara frigorífica**	volume utile d'une chambre froide	5.3-**11**
freezer room ● *frozen-food storage room*	**cámara (de almacenamiento) para productos congelados**	chambre (d'entreposage) pour produits congelés	5.3-**12**
frozen-food storage room *freezer room* ●	**cámara (de almacenamiento) para productos congelados**	chambre (d'entreposage) pour produits congelés	5.3-**13**
gross capacity *gross volume* ●	**volumen total de una cámara frigorífica** volumen bruto de una cámara frigorífica	volume brut d'une chambre froide	5.3-**14**
gross volume ● *gross capacity*	**volumen total de una cámara frigorífica** volumen bruto de una cámara frigorífica	volume brut d'une chambre froide	5.3-**15**
inflatable cold store	**cámara frigorífica inflable**	chambre froide gonflable	5.3-**16**
jacketed cold room	**cámara frigorífica de doble pared**	chambre (froide) à double paroi *chambre (froide) à enveloppe d'air froid*	5.3-**17**

ENGLISH	ESPAÑOL	FRANÇAIS
5.3-**18** net capacity *effective capacity* ○	**volumen útil de una cámara frigorífica**	volume utile d'une chambre froide
5.3-**19** portable cold room	**cámara frigorífica desmontable**	chambre froide transportable
5.3-**20** precooling room	**cámara de prerrefrigeración**	chambre de préréfrigération
5.3-**21** sectional cold room	**cámara frigorífica prefabricada**	chambre froide préfabriquée
5.3-**22** store contents	**existencias** carga	chargement d'une chambre
5.3-**23** vestibule with doors	**manga** túnel	antichambre
5.3-**24** walk-in freezer ○	**pequeña cámara para productos congelados**	congélateur-chambre ○
5.3-**25** walk-in refrigerator ○	**cámara frigorífica pequeña**	réfrigérateur-chambre ○

SECTION 5.4 *Cold stores*	SUBCAPÍTULO 5.4 *Almacenes frigoríficos*	SOUS-CHAPITRE 5.4 *Entrepôts frigorifiques*
5.4-**1** bulk storage *storage in bulk* *bulk stowage* ○	**almacenamiento a granel**	entreposage en vrac
5.4-**2** bulk stowage ○ *bulk storage* *storage in bulk*	**almacenamiento a granel**	entreposage en vrac
5.4-**3** cold storage *refrigerated storage*	**almacenamiento frigorífico**	entreposage frigorifique
5.4-**4** cold store *refrigerated warehouse*	**almacén frigorífico**	entrepôt frigorifique
5.4-**5** cold store combine ○ *cold store facility* *cold store complex* ○	**complejo frigorífico**	complexe frigorifique
5.4-**6** cold store complex ○ *cold store facility* *cold store combine* ○	**complejo frigorífico**	complexe frigorifique
5.4-**7** cold store facility *cold store combine* ○ *cold store complex* ○	**complejo frigorífico**	complexe frigorifique
5.4-**8** despatching cold store	**almacén frigorífico de distribución**	entrepôt frigorifique de distribution
5.4-**9** dock leveller	**dispositivo para puesta a nivel (respecto al del muelle)**	dispositif de mise à niveau (de quai)
5.4-**10** engine house	**edificio de maquinaria**	bâtiment des machines
5.4-**11** frozen-food store	**almacén para productos congelados**	entrepôt pour produits congelés
5.4-**12** gangway ○ *passage*	**pasillo**	allée *passage* ○
5.4-**13** high-rise store	**almacén con cámara de gran altura**	entrepôt de grande hauteur
5.4-**14** loading bank ○ *platform*	**muelle** andén	quai
5.4-**15** maximum stacking density	**densidad máxima de estiba** densidad máxima de apilado	densité de gerbage
5.4-**16** multistorey cold store ○	**almacén frigorífico de varias plantas**	entrepôt frigorifique à étages

ENGLISH	ESPAÑOL	FRANÇAIS	
multipurpose cold store	**almacén frigorífico polivalente**	entrepôt frigorifique polyvalent	5.4-**17**
nursery cold store	**cámara frigorífica para plantas de vivero**	chambre froide à plants	5.4-**18**
order-picking area	**zona de preparación de pedidos**	zone de préparation des commandes	5.4-**19**
palletizing area	**área de manipulación con tarimas**	aire de palettisation	5.4-**20**
passage *gangway* ○	**pasillo**	allée *passage* ◑	5.4-**21**
platform *loading bank* ○	**muelle** andén	quai	5.4-**22**
platform height floor *truck height floor* ◑	**piso bajo a nivel del muelle**	plancher bas à niveau de quai	5.4-**23**
port cold store	**almacén frigorífico portuario**	entrepôt frigorifique portuaire	5.4-**24**
port door	**puerta de atraque**	porte d'accostage	5.4-**25**
private cold store	**almacén frigorífico privado**	entrepôt frigorifique privé	5.4-**26**
public cold store	**almacén frigorífico público**	entrepôt frigorifique public	5.4-**27**
ramp *steep* ◑	**rampa de acceso**	rampe d'accès	5.4-**28**
refrigerated storage *cold storage*	**almacenamiento frigorífico**	entreposage frigorifique	5.4-**29**
refrigerated warehouse *cold store*	**almacén frigorífico**	entrepôt frigorifique	5.4-**30**
shelf storage *shelf stowage* ◑	**almacenamiento en estantes**	entreposage sur étagères	5.4-**31**
shelf stowage ◑ *shelf storage*	**almacenamiento en estantes**	entreposage sur étagères	5.4-**32**
single-storey cold store ◑	**almacén frigorífico de una sola planta**	entrepôt frigorifique à un seul niveau	5.4-**33**
slab-on-ground floor	**piso bajo al nivel del suelo**	plancher bas à niveau de sol	5.4-**34**
specialized cold store	**almacén frigorífico especializado**	entrepôt frigorifique spécialisé	5.4-**35**
stacking allotment	**distribución en lotes**	allotissement *lotissement*	5.4-**36**
stacking density	**densidad de almacenamiento**	densité maximale d'entreposage	5.4-**37**
steep ◑ *ramp*	**rampa de acceso**	rampe d'accès	5.4-**38**
stock turnover	**rotación de los productos almacenados**	rotation des stocks	5.4-**39**
storage factor	**coeficiente de ocupación (de un almacén)** coeficiente de utilización (de un almacén)	taux de remplissage (d'un entrepôt)	5.4-**40**
storage in bulk *bulk storage* *bulk stowage* ◑	**almacenamiento a granel**	entreposage en vrac	5.4-**41**
store *warehouse*	**almacén** depósito	entrepôt	5.4-**42**
truck height floor ◑ *platform height floor*	**piso bajo a nivel del muelle**	plancher bas à niveau de quai	5.4-**43**
warehouse *store*	**almacén** depósito	entrepôt	5.4-**44**
warming room	**sala de acondicionamiento térmico** *cámara de recalentamiento*	salle de réchauffage	5.4-**45**

	ENGLISH	ESPAÑOL	FRANÇAIS
	SECTION 5.5 *Refrigerated transport* **SECTION 5.5.1** *Refrigerated transport special technology*	**SUBCAPÍTULO 5.5** *Transporte frigorífico* **SUBCAPÍTULO 5.5.1** *Técnicas específicas del transporte frigorífico*	**SOUS-CHAPITRE 5.5** *Transport frigorifique* **SOUS-CHAPITRE 5.5.1** *Techniques spécifiques au transport frigorifique*
5.5.1-1	aeolian fan *wind-turbine fan*	**ventilador de turbina eólica**	ventilateur à turbine éolienne
5.5.1-2	air leakage	**fuga de aire**	fuite d'air
5.5.1-3	ATP agreement	**acuerdo ATP**	accord ATP
5.5.1-4	body-icing	**aplicación directa de hielo a la carga**	glaçage direct au sein du chargement
5.5.1-5	combined transport	**transporte combinado**	transport combiné
5.5.1-6	contact icing	**aplicación directa de hielo (a los productos)**	glaçage direct (des denrées)
5.5.1-7	controlled-atmosphere transport	**transporte en atmósfera controlada**	transport sous atmosphère contrôlée
5.5.1-8	controlled-temperature transport	**transporte a temperatura controlada** transporte a temperatura regulada	transport sous température dirigée
5.5.1-9	dry-ice bunker	**compartimiento de nieve carbónica**	compartiment à glace carbonique
5.5.1-10	expendable refrigerant system	**sistema con frigorígeno perdido**	équipement à frigorigène perdu
5.5.1-11	ice blower *ice gun*	**lanzahielo**	lance-glace
5.5.1-12	ice bunker	**depósito para hielo**	bac à glace *panier à glace*
5.5.1-13	ice-bunker screen	**pantalla de depósito de hielo**	écran de panier à glace
5.5.1-14	ice gun *ice blower*	**lanzahielo**	lance-glace
5.5.1-15	icing hatch	**trampilla para la carga del hielo**	trappe de chargement de la glace
5.5.1-16	insulated body	**caja aislada** caja isoterma caja isotérmica	caisse isolée *caisse isotherme* ○
5.5.1-17	insulated vehicle	**vehículo isotermo**	véhicule isolé *véhicule isotherme* ○
5.5.1-18	interface	**interfaz**	interface
5.5.1-19	intermodal transport	**transporte combinado**	transport intermodal
5.5.1-20	land transport equipment	**vehículo de transporte terrestre**	engin de transport terrestre
5.5.1-21	mechanically refrigerated vehicle	**vehículo frigorífico**	véhicule frigorifique
5.5.1-22	multimodal transport	**transporte multimodal**	transport multimodal
5.5.1-23	over-the-road system ○	**sistema de enfriamiento en ruta**	système de refroidissement en route ○
5.5.1-24	piggy-back transport	**transporte combinado por ferrocarril y carretera**	ferroutage *transport rail-route* ○
5.5.1-25	re-icing	**repostado de hielo**	reglaçage
5.5.1-26	refrigerated transport	**transporte frigorífico**	transport frigorifique
5.5.1-27	refrigerated vehicle	**vehículo refrigerante** vehículo refrigerado	véhicule réfrigérant
5.5.1-28	top-icing	**aplicación directa de hielo sobre la carga** recubrimiento (de la carga) con hielo triturado	glaçage direct sur le chargement

ENGLISH	ESPAÑOL	FRANÇAIS	
ventilated vehicle	**vehículo ventilado**	véhicule aéré	5.5.1-**29**
wind-turbine fan *aeolian fan*	**ventilador de turbina eólica**	ventilateur à turbine éolienne	5.5.1-**30**

SECTION 5.5.2 *Road transport*	SUBCAPÍTULO 5.5.2 *Transporte por carretera*	SOUS-CHAPITRE 5.5.2 *Transport routier*	
air-flow floor *rack floor*	**piso ventilado** piso enjaretado	plancher soufflant *plancher ventilé*	5.5.2-**1**
curtain-sider	**remolque con laterales desmontables**	remorque à bâches latérales amovibles	5.5.2-**2**
delivery vehicle	**vehículo isotermo de reparto**	véhicule de livraison isotherme	5.5.2-**3**
distribution hub	**plataforma de distribución**	plate-forme de distribution	5.5.2-**4**
fuel-freeze system	**sistema dual de enfriamiento y carburante**	équipement à frigorigène courant	5.5.2-**5**
insulated large van *insulated lorry* *insulated truck (USA)*	**camión isotermo**	camion isolé	5.5.2-**6**
insulated lorry *insulated truck (USA)* *insulated large van*	**camión isotermo**	camion isolé	5.5.2-**7**
insulated truck (USA) *insulated lorry* *insulated large van*	**camión isotermo**	camion isolé	5.5.2-**8**
mechanically refrigerated large van *mechanically refrigerated lorry* *mechanically refrigerated truck (USA)*	**camión frigorífico**	camion frigorifique	5.5.2-**9**
mechanically refrigerated lorry *mechanically refrigerated large van* *mechanically refrigerated truck (USA)*	**camión frigorífico**	camion frigorifique	5.5.2-**10**
mechanically refrigerated truck (USA) *mechanically refrigerated lorry* *mechanically refrigerated large van*	**camión frigorífico**	camion frigorifique	5.5.2-**11**
multicompartment vehicle	**vehículo compartimentalizado**	véhicule multicompartiment	5.5.2-**12**
nose-mounted refrigeration unit	**grupo frigorífico frontal**	groupe frigorifique frontal	5.5.2-**13**
rack floor *air-flow floor*	**piso ventilado** piso enjaretado	plancher soufflant *plancher ventilé*	5.5.2-**14**
refrigerated large van *refrigerated lorry* *refrigerated truck (USA)*	**camión refrigerante** camión refrigerado	camion réfrigérant	5.5.2-**15**
refrigerated lorry *refrigerated large van* *refrigerated truck (USA)*	**camión refrigerante** camión refrigerado	camion réfrigérant	5.5.2-**16**
refrigerated truck (USA) *refrigerated lorry* *refrigerated large van*	**camión refrigerante** camión refrigerado	camion réfrigérant	5.5.2-**17**
road tanker ❍ *tank truck (USA)* ❍	**camión cisterna**	camion citerne	5.5.2-**18**
tank truck (USA) ❍ *road tanker* ❍	**camión cisterna**	camion citerne	5.5.2-**19**
trailer	**remolque**	remorque	5.5.2-**20**
underslung refrigeration unit	**grupo frigorífico bajo chasis**	groupe frigorifique sous châssis	5.5.2-**21**

ENGLISH	ESPAÑOL	FRANÇAIS
SECTION 5.5.3 *Rail transport*	**SUBCAPÍTULO 5.5.3** *Transporte por ferrocarril*	**SOUS-CHAPITRE 5.5.3** *Transport ferroviaire*
5.5.3-**1** block train	**convoy frigorífico** tren frigorífico	train bloc
5.5.3-**2** end-bunker refrigerated railcar (USA) ◐ *end-bunker refrigerated truck* ◐	**vagón refrigerante con depósitos de hielo en los testeros** vagón refrigerado con depósitos de hielo en los testeros	wagon réfrigérant à bac d'extrémité ◐
5.5.3-**3** end-bunker refrigerated truck ◐ *end-bunker refrigerated railcar (USA)* ◐	**vagón refrigerante con depósitos de hielo en los testeros** vagón refrigerado con depósitos de hielo en los testeros	wagon réfrigérant à bac d'extrémité ◐
5.5.3-**4** iced railcar (USA) *iced truck*	**vagón refrigerante con hielo** vagón refrigerado con hielo	wagon glacé
5.5.3-**5** iced truck *iced railcar (USA)*	**vagón refrigerante con hielo** vagón refrigerado con hielo	wagon glacé
5.5.3-**6** icing of railcars (USA) *icing of trucks*	**carga de hielo en los vagones**	glaçage des wagons
5.5.3-**7** icing of trucks *icing of railcars (USA)*	**carga de hielo en los vagones**	glaçage des wagons
5.5.3-**8** icing platform	**plataforma de carga de hielo**	plate-forme de glaçage
5.5.3-**9** icing tower	**torre de carga de hielo**	tour de glaçage
5.5.3-**10** insulated railcar (USA) *insulated railway truck* *insulated truck*	**vagón isotermo**	wagon isolé
5.5.3-**11** insulated railway truck *insulated truck* *insulated railcar (USA)*	**vagón isotermo**	wagon isolé
5.5.3-**12** insulated truck *insulated railway truck* *insulated railcar (USA)*	**vagón isotermo**	wagon isolé
5.5.3-**13** mechanically refrigerated railcar (USA) *mechanically refrigerated truck* *refrigerator car (USA)*	**vagón frigorífico**	wagon frigorifique
5.5.3-**14** mechanically refrigerated truck *mechanically refrigerated railcar (USA)* *refrigerator car (USA)*	**vagón frigorífico**	wagon frigorifique
5.5.3-**15** overhead-bunker refrigerated railcar (USA) *overhead-bunker refrigerated truck*	**vagón refrigerante con depósitos de hielo en el techo** vagón refrigerado con depósitos de hielo en el techo	wagon réfrigérant à bac plafonnier
5.5.3-**16** overhead-bunker refrigerated truck *overhead-bunker refrigerated railcar (USA)*	**vagón refrigerante con depósitos de hielo en el techo** vagón refrigerado con depósitos de hielo en el techo	wagon réfrigérant à bac plafonnier
5.5.3-**17** rail tanker ◐	**vagón cisterna**	wagon citerne ◐
5.5.3-**18** refrigerated railcar (USA) *refrigerated truck*	**vagón refrigerante** vagón refrigerado	wagon réfrigérant
5.5.3-**19** refrigerated train	**tren frigorífico**	train frigorifique
5.5.3-**20** refrigerated truck *refrigerated railcar (USA)*	**vagón refrigerante** vagón refrigerado	wagon réfrigérant

ENGLISH	ESPAÑOL	FRANÇAIS	
refrigerator car (USA) *mechanically refrigerated truck* *mechanically refrigerated railcar (USA)*	**vagón frigorífico**	wagon frigorifique	5.5.3-**21**

SECTION 5.5.4 *Marine transport*	SUBCAPÍTULO 5.5.4 *Transporte marítimo*	SOUS-CHAPITRE 5.5.4 *Transport maritime*	
barge carrier	**barco portabarcazas** barco portagabarras	navire porte-barges *navire porte-chalands*	5.5.4-**1**
bottom plating	**forro del fonde** vagras del fondo	vaigrage de fond	5.5.4-**2**
bulkhead	**mampara**	cloison	5.5.4-**3**
cargo door *hull loading door*	**porta de carga** portalón de carga	portelone *sabord de charge*	5.5.4-**4**
cells (of a container ship)	**celdas (de barco portacontenedores)**	cellules (de navires porte-conteneurs)	5.5.4-**5**
cellular (container) ship	**barco (portacontenedores) celular**	navire (porte-conteneurs) cellulaire	5.5.4-**6**
(container) guides	**guías para contenedores**	guides (de conteneurs)	5.5.4-**7**
container ship *container vessel*	**barco portacontenedores**	navire porte-conteneurs	5.5.4-**8**
container vessel *container ship*	**barco portacontenedores**	navire porte-conteneurs	5.5.4-**9**
deck	**cubierta** puente	pont	5.5.4-**10**
dry cargo	**mercancías no enfriadas**	marchandises ordinaires	5.5.4-**11**
factory ship	**barco fábrica** buque factoría	navire-usine	5.5.4-**12**
freezing trawler	**(barco) arrastrero congelador**	chalutier congélateur	5.5.4-**13**
gas-carrier	**barco para transporte de gases**	navire gazier	5.5.4-**14**
hatch ○	**cuartel (de escotilla)** tapa de escotilla	écoutille *panneau (de cale)* ○	5.5.4-**15**
hold	**bodega**	cale	5.5.4-**16**
hold pen ○ *hold pound*	**compartimiento de bodega**	compartiment de cale	5.5.4-**17**
hold pound *hold pen* ○	**compartimiento de bodega**	compartiment de cale	5.5.4-**18**
hull	**casco**	coque	5.5.4-**19**
hull loading door *cargo door*	**porta de carga** portalón de carga	portelone *sabord de charge*	5.5.4-**20**
hull plating *planking* *shell plating*	**revestimiento del casco**	bordé extérieur	5.5.4-**21**
inner lining *inner planking* ○ *inner plating* ○	**forro interior** revestimiento interior (barcos)	vaigrage	5.5.4-**22**
inner planking ○ *inner lining* *inner plating* ○	**forro interior** revestimiento interior (barcos)	vaigrage	5.5.4-**23**
inner plating ○ *inner lining* *inner planking* ○	**forro interior** revestimiento interior (barcos)	vaigrage	5.5.4-**24**

	ENGLISH	ESPAÑOL	FRANÇAIS
5.5.4-**25**	integrated tank ○ *membrane tank* ○	**tanque integrado** tanque de membrana	cuve à membrane *cuve intégrée*
5.5.4-**26**	marine refrigerating plant	**instalación frigorífica marina** instalación frigorífica a bordo	installation frigorifique marine
5.5.4-**27**	membrane tank ○ *integrated tank* ○	**tanque integrado** tanque de membrana	cuve à membrane *cuve intégrée*
5.5.4-**28**	methane carrier *methane tanker*	**(barco) metanero**	(navire) méthanier
5.5.4-**29**	methane tanker *methane carrier*	**(barco) metanero**	(navire) méthanier
5.5.4-**30**	mother ship	**barco nodriza** buque nodriza	navire-base
5.5.4-**31**	planking *hull plating* *shell plating*	**revestimiento del casco**	bordé extérieur
5.5.4-**32**	reefer ship *refrigerated cargo vessel*	**barco frigorífico** buque frigorífico	navire frigorifique
5.5.4-**33**	refrigerated (cargo) hold	**bodega refrigerada**	cale refroidie
5.5.4-**34**	refrigerated cargo vessel *reefer ship*	**barco frigorífico** buque frigorífico	navire frigorifique
5.5.4-**35**	roll-on roll-off ship	**barco para cargas sobre ruedas**	roulier *transroulier* ○
5.5.4-**36**	self-supporting tank ○ *structural tank*	**tanque autoportante**	cuve autoporteuse
5.5.4-**37**	shell plating *hull plating* *planking*	**revestimiento del casco**	bordé extérieur
5.5.4-**38**	stanchion	**puntal**	épontille ○
5.5.4-**39**	structural tank *self-supporting tank* ○	**tanque autoportante**	cuve autoporteuse
5.5.4-**40**	tweendeck	**entrepuente**	entrepont

	SECTION 5.5.5 *Air transport*	SUBCAPÍTULO 5.5.5 *Transporte aéreo*	SOUS-CHAPITRE 5.5.5 *Transport aérien*
5.5.5-**1**	air cargo	**cargamento** flete	fret aérien
5.5.5-**2**	air carrier	**transportador aéreo** operador aéreo	transporteur aérien
5.5.5-**3**	air container *ULD (Unit Load Device)*	**ULD (unidad de carga)**	ULD (unité de charge) *conteneur aérien*
5.5.5-**4**	air freight	**flete aéreo**	fret aérien
5.5.5-**5**	anti-icer	**dispositivo antihielo**	anti-givre
5.5.5-**6**	anti-icing	**antihielo**	anti-givrage
5.5.5-**7**	belly cargo *belly hold*	**bodega**	fret soute inférieure
5.5.5-**8**	belly hold *belly cargo*	**bodega**	fret soute inférieure
5.5.5-**9**	clearance	**espacio libre**	espace libre

ENGLISH	ESPAÑOL	FRANÇAIS	
combi aircraft	**avión de pasaje y carga**	avion combi	5.5.5-**10**
ground equipment	**equipamiento en tierra**	équipement au sol	5.5.5-**11**
hub	**aeropuerto base**	aéroport principal *plaque tournante*	5.5.5-**12**
igloo	**contenedor igloo**	conteneur igloo	5.5.5-**13**
insulating ULD *thermal ULD*	**ULD aislado** ULD térmico	unité de charge isolée	5.5.5-**14**
landing weight	**tara al aterrizaje**	masse à l'atterrissage	5.5.5-**15**
loose cargo	**carga a granel** flete a granel	fret en vrac	5.5.5-**16**
lower deck *lower hold*	**puente inferior** bodega inferior	pont inférieur	5.5.5-**17**
lower hold *lower deck*	**puente inferior** bodega inferior	pont inférieur	5.5.5-**18**
main deck	**puente principal** bodega principal	pont principal	5.5.5-**19**
PAG *pallet*	**pallet**	palette	5.5.5-**20**
pallet *PAG*	**pallet**	palette	5.5.5-**21**
payload	**carga transportada**	charge marchande	5.5.5-**22**
perishable cargo	**carga perecedera**	fret périssable	5.5.5-**23**
pressurization	**presurización**	pressurisation	5.5.5-**24**
ramp weight	**peso bruto**	masse au stationnement	5.5.5-**25**
thermal ULD *insulating ULD*	**ULD aislado** ULD térmico	unité de charge isolée	5.5.5-**26**
turnaround	**tiempo de espera**	temps nécessaire au déchargement et au chargement	5.5.5-**27**
ULD (Unit Load Device) *air container*	**ULD (unidad de carga)**	ULD (unité de charge) *conteneur aérien*	5.5.5-**28**
widebody aircraft	**avión de gran capacidad**	avion gros porteur	5.5.5-**29**

SECTION 5.6 *Handling equipment*	SUBCAPÍTULO 5.6 *Material de manipulación*	SOUS-CHAPITRE 5.6 *Matériel de manutention*	
box pallet *bulk bin* *pallet bin* *pallet box* *pallet crate* *palletainer* *pallox*	**caja-palet** jaula-tarima	caisse palette *pallox* ◐	5.6-**1**
bulk bin *box pallet* *pallet bin* *pallet box* *pallet crate* *palletainer* *pallox*	**caja-palet** jaula-tarima	caisse palette *pallox* ◐	5.6-**2**
bulkhead	**mampara pantalla**	cloison "écran"	5.6-**3**

	ENGLISH	ESPAÑOL	FRANÇAIS
5.6-**4**	bulking *grouping*	**agrupación de cargas**	groupage
5.6-**5**	ceiling air duct	**conducto de aire frío por el techo**	conduit de ventilation au plafond
5.6-**6**	clip-on unit	**grupo frigorífico movil** grupo frigorífico portable	groupe frigorifique amovible
5.6-**7**	converter bar *pallet converter* *pallet posts*	**jaula (para tarimas)**	convertisseur (pour palettes)
5.6-**8**	depalletize (to)	**despaletizar**	dépalettiser
5.6-**9**	dolly	**soporte con rodillos**	support à roulettes
5.6-**10**	europallet	**europalet**	europalette
5.6-**11**	floor air duct	**conducto de aire frío por el suelo**	conduit sous plancher
5.6-**12**	forklift truck	**carretilla eleadora de horquilla**	chariot (élévateur) à fourche
5.6-**13**	(freight) container *(transport) container*	**contenedor**	conteneur (de transport) *container* ○
5.6-**14**	grouping *bulking*	**agrupación de cargas**	groupage
5.6-**15**	handling	**manipulación**	manutention
5.6-**16**	heated container	**contenedor calefactado**	conteneur chauffé
5.6-**17**	insulated container	**contenedor aislado**	conteneur isolé
5.6-**18**	ISO container	**contenedor ISO**	conteneur ISO
5.6-**19**	lift truck	**carretilla elevadora**	chariot élévateur
5.6-**20**	mechanically refrigerated container	**contenedor frigorífico**	conteneur frigorifique
5.6-**21**	pallet	**tarima**	palette
5.6-**22**	pallet bin *box pallet* *bulk bin* *pallet box* *pallet crate* *palletainer* *pallox*	**caja-palet** jaula-tarima	caisse palette *pallox* ◐
5.6-**23**	pallet box *box pallet* *bulk bin* *pallet bin* *pallet crate* *palletainer* *pallox*	**caja-palet** jaula-tarima	caisse palette *pallox* ◐
5.6-**24**	pallet converter *converter bar* *pallet posts*	**jaula (para tarimas)**	convertisseur (pour palettes)
5.6-**25**	pallet crate *box pallet* *bulk bin* *pallet bin* *pallet box* *palletainer* *pallox*	**caja-palet** jaula-tarima	caisse palette *pallox* ◐
5.6-**26**	pallet loader *pallet truck* *pedestrian forklift*	**carretilla manual transportadora de tarimas**	transpalette
5.6-**27**	pallet posts *converter bar* *pallet converter*	**jaula (para tarimas)**	convertisseur (pour palettes)

ENGLISH	ESPAÑOL	FRANÇAIS	
pallet rack	**estanteria para tarimas**	palettier	5.6-**28**
pallet truck *pallet loader* *pedestrian forklift*	**carretilla manual transportadora de tarimas**	transpalette	5.6-**29**
palletainer *box pallet* *bulk bin* *pallet bin* *pallet box* *pallet crate* *pallox*	**caja-palet** jaula-tarima	caisse palette *pallox* ○	5.6-**30**
palletization	**manipulación con tarimas**	palettisation	5.6-**31**
palletize (to)	**manipular con tarimas**	palettiser	5.6-**32**
pallox *box pallet* *bulk bin* *pallet bin* *pallet box* *pallet crate* *palletainer*	**caja-palet** jaula-tarima	caisse palette *pallox* ○	5.6-**33**
pedestrian forklift *pallet loader* *pallet truck*	**carretilla manual transportadora de tarimas**	transpalette	5.6-**34**
pile (to) *stack (to)*	**estibar** apilar	gerber *empiler*	5.6-**35**
porthole container	**contenedor con portillas**	conteneur à hublots	5.6-**36**
reach truck	**carretilla elevadora de horquilla retráctil**	chariot (élévateur) à fourche rétractable *chariot élévateur à mât rétractable* ○	5.6-**37**
reefer container *refrigerated container*	**contenedor frigorífico**	conteneur réfrigéré	5.6-**38**
refrigerated container *reefer container*	**contenedor frigorífico**	conteneur réfrigéré	5.6-**39**
refrigerated and heated container	**contenedor refrigerado y calefactado**	conteneur réfrigéré et chauffé	5.6-**40**
refrigerated and heated container with controlled or modified atmosphere	**contenedor de atmósfera controlada**	conteneur réfrigéré et chauffé à atmosphère contrôlée	5.6-**41**
removable equipment	**equipo móvil** equipo portable	équipement amovible *unité amovible* ○	5.6-**42**
roll container	**contenedor con ruedas**	roll conteneur	5.6-**43**
stack (to) *pile (to)*	**estibar** apilar	gerber *empiler*	5.6-**44**
stack height	**altura de estiba** altura de apilamiento	hauteur de gerbage	5.6-**45**
stacking	**estiba**	gerbage	5.6-**46**
swap body	**contenedor intercambiable (ferrocarril/ carretera)**	conteneur rail-route	5.6-**47**
tank container	**contenedor cisterna**	conteneur-citerne	5.6-**48**
test pressure	**test de presión** ensayo de presión	pression d'épreuve	5.6-**49**
thermal container	**contenedor isotermo**	conteneur à caractéristiques thermiques	5.6-**50**
(transport) container *(freight) container*	**contenedor**	conteneur (de transport) *container* ○	5.6-**51**

Capítulo 6.

Aplicación del frío a los productos perecederos

● término aceptado

○ término obsoleto

ENGLISH	ESPAÑOL	FRANÇAIS	
SECTION 6.1 *Food characteristics*	**SUBCAPÍTULO 6.1** *Características de los alimentos*	**SOUS-CHAPITRE 6.1** *Caractéristiques des aliments*	
aroma	**aroma**	arôme *parfum* ○	6.1-**1**
boil-in-bag food	**alimento precocinado en bolsas**	aliment précuit en sachet	6.1-**2**
bound water	**agua ligada**	eau liée	6.1-**3**
chilled food	**alimento refrigerado**	denrées réfrigérées	6.1-**4**
cloudiness	**turbidez**	turbidité *louche* ○ *trouble* ○	6.1-**5**
convenience food	**alimento listo para su empleo** alimento listo para comer	aliment prêt à l'emploi	6.1-**6**
fermentation heat	**calor de fermentación**	chaleur de fermentation	6.1-**7**
flavour	**gusto**	flaveur	6.1-**8**
food	**alimento**	aliment	6.1-**9**
free water	**agua libre**	eau libre	6.1-**10**
freeze-thaw resistance	**resistencia a la congelación-descongelación**	résistance à la congélation-décongélation	6.1-**11**
institutional food	**alimento para colectividades**	aliment pour collectivité	6.1-**12**
keeping quality	**conservabilidad** aptitud para la conservación	aptitude à la conservation	6.1-**13**
keeping time ○ *storage time*	**duración potencial de almacenamiento** vida potencial de almacenamiento periodo de conservación	durée de conservation *durée d'entreposage* ○	6.1-**14**
metabolic heat	**calor metabólico**	chaleur métabolique	6.1-**15**
minimally prepared food	**producto alimentario de 4ª gama**	produit alimentaire de 4e gamme	6.1-**16**
organoleptic qualities	**cualidades organolépticas**	qualités organoleptiques *qualités sensorielles* ○	6.1-**17**
practical storage life	**duración práctica de conservación**	durée pratique de conservation *durée pratique d'entreposage*	6.1-**18**
precooked dish ○ *prepared food*	**alimento precocinado** plato preparado	aliment cuisiné *aliment précuit* *plat préparé* ○	6.1-**19**
prepared food *precooked dish* ○	**alimento precocinado** plato preparado	aliment cuisiné *aliment précuit* *plat préparé* ○	6.1-**20**
quality assessment *quality evaluation* ○	**apreciación de la calidad** evaluación de la calidad	évaluation de la qualité *appréciation de la qualité* ○	6.1-**21**
quality control	**control de la calidad**	contrôle de la qualité *maintien de la qualité* *maîtrise de la qualité*	6.1-**22**
quality evaluation ○ *quality assessment*	**apreciación de la calidad** evaluación de la calidad	évaluation de la qualité *appréciation de la qualité* ○	6.1-**23**
rate of respiration *respiratory intensity* ○	**intensidad respiratoria**	intensité respiratoire	6.1-**24**
respiratory heat	**calor de respiración**	chaleur de respiration	6.1-**25**
respiratory intensity ○ *rate of respiration*	**intensidad respiratoria**	intensité respiratoire	6.1-**26**

	ENGLISH	ESPAÑOL	FRANÇAIS
6.1-27	sensory (organoleptic) properties	**propiedades organolépticas** propiedades sensoriales	propriétés organoleptiques *propriétés sensorielles* ◖
6.1-28	storage time *keeping time* ○	**duración potencial de almacenamiento** vida potencial de almacenamiento periodo de conservación	durée de conservation *durée d'entreposage* ◖
6.1-29	sugar content	**contenido en azúcar**	teneur en sucre
6.1-30	taste	**sabor**	goût
6.1-31	time-temperature tolerance (TTT)	**tolerancia tiempo-temperatura**	corrélation temps-température *tolérance temps-température* ◖
6.1-32	tolerance	**tolerancia**	tolérance

	SECTION 6.2 *Food deterioration and hygiene*	SUBCAPÍTULO 6.2 *Alteraciones de los alimentos e higiene alimentaria*	SOUS-CHAPITRE 6.2 *Altération des aliments et hygiène alimentaire*
6.2-1	acceptance criteria	**criterio de aceptación**	critère d'acceptabilité
6.2-2	acceptance level	**nivel de aceptación**	niveau d'acceptabilité
6.2-3	adverse influence	**deterioro** alteración	altération *détérioration* ◖
6.2-4	bacterial decay ◖ *bacterial deterioration*	**alteración bacteriano** deterioro bacteriano	altération bactérienne
6.2-5	bacterial deterioration *bacterial decay* ◖	**alteración bacteriano** deterioro bacteriano	altération bactérienne
6.2-6	chilling injury	**daño por frío**	accident causé par la réfrigération
6.2-7	clean	**sano** limpio	propre
6.2-8	cleanable	**limpiable** desinfectable	nettoyable
6.2-9	cleaning	**limpieza** desinfección	nettoyage
6.2-10	cold injury *low-temperature injury* ◖	**daño por frío o congelación**	altération due aux basses températures
6.2-11	contaminant	**contaminante**	contaminant
6.2-12	contamination	**contaminación**	contamination
6.2-13	control	**control**	maîtrise
6.2-14	control measures	**medidas de control**	mesures de maîtrise
6.2-15	control (to)	**controlar**	maîtriser
6.2-16	control point	**punto de control**	point de maîtrise
6.2-17	corrective action	**acción correctora**	action corrective
6.2-18	corrosion-resistant material	**material anticorrosivo**	matériau résistant à la corrosion
6.2-19	crevice	**grieta** hendidura	crevasse *anfractuosité* ◖
6.2-20	Critical Control Point (CCP)	**punto de control crítico**	point critique pour la maîtrise (CCP)
6.2-21	critical limit	**límite crítico**	limite critique
6.2-22	cryophilic ○	**criófilo**	cryophile

ENGLISH	ESPAÑOL	FRANÇAIS	
dead space	**espacio muerto**	zone non accessible	6.2-**23**
deterioration *spoilage*	**deterioro** alteración	altération *détérioration* ○	6.2-**24**
discolouration	**alteración del color**	altération de la couleur *décoloration*	6.2-**25**
disinfection	**desinfección**	désinfection	6.2-**26**
dose-response assessment	**evaluación dosis-respuesta**	appréciation de la relation dose-réponse	6.2-**27**
durable	**durabilidad**	résistante	6.2-**28**
enzymatic reactions	**reacciones enzimáticas**	réactions enzymatiques	6.2-**29**
exposure assessment	**evaluación de riesgos** evaluación de peligros	évaluation de l'exposition (à des agents indésirables)	6.2-**30**
food area	**superficie del alimento**	zone alimentaire	6.2-**31**
food hygiene	**higiene alimentaria**	hygiène des aliments *hygiène alimentaire*	6.2-**32**
food safety	**seguridad alimentaria**	sécurité sanitaire des aliments	6.2-**33**
food suitability	**adecuación del alimento**	acceptabilité des aliments	6.2-**34**
freezer burn	**quemadura superficial por congelación**	brûlure de congélation	6.2-**35**
freezing damage	**daño por congelación**	altération due à la congélation	6.2-**36**
freezing injury	**herida por congelación** daño por congelación	altération au cours de la congélation ou de l'entreposage (du produit congelé)	6.2-**37**
frost damage	**herida por congelación** daño por congelación	gelure	6.2-**38**
gas injury	**daño por gas**	intoxication gazeuse	6.2-**39**
HACCP (Hazard Analysis Critical Control Point)	**APPCC (Análisis de Peligros y Puntos de Control Critico)**	HACCP (analyse des dangers et leur maîtrise aux points critiques)	6.2-**40**
HACCP plan	**plan del APPCC**	plan HACCP	6.2-**41**
hazard	**riesgo** peligro	danger	6.2-**42**
hazard analysis	**análisis de riesgos** análisis de peligros	analyse des dangers	6.2-**43**
hazard assessment	**evaluación de riesgos** evaluación de peligros	appréciation des dangers	6.2-**44**
hazard identification	**identificación de riesgos** identificación de peligros	identification des dangers	6.2-**45**
in-package desiccation	**desecación superficial de un producto congelado y envasado**	dessiccation dans les emballages	6.2-**46**
inspection	**inspección**	contrôle	6.2-**47**
low-temperature injury ○ *cold injury*	**daño por frío o congelación**	altération due aux basses températures	6.2-**48**
mechanical defect	**daño mecánico**	altération mécanique *blessure*	6.2-**49**
moisture loss	**deshidratación**	perte d'eau	6.2-**50**
monitoring	**vigilancia**	surveillance	6.2-**51**
non-absorbent material	**material no absorbente**	matériau non absorbant	6.2-**52**
non-food area	**área no alimentaria**	zone non alimentaire	6.2-**53**

	ENGLISH	ESPAÑOL	FRANÇAIS
6.2-**54**	non-toxic material	**material no tóxico**	matériau non toxique
6.2-**55**	oxidation	**oxidación**	oxydation
6.2-**56**	perishable	**perecedero**	périssable
6.2-**57**	preventive action	**acción preventiva**	action préventive *prévention*
6.2-**58**	procedure	**procedimiento**	procédure
6.2-**59**	psychrophilic	**psicrófilo**	psychrophile
6.2-**60**	psychrotrophic	**psicrotrofo**	psychrotrophe
6.2-**61**	quality audit	**auditoría de calidad**	audit de qualité
6.2-**62**	quality control	**control de calidad**	maîtrise de la qualité
6.2-**63**	quality insurance	**aseguramiento de la calidad**	assurance qualité
6.2-**64**	quality surveillance	**vigilancia de calidad** supervisión de calidad	surveillance de la qualité *contrôle de la qualité*
6.2-**65**	rancidity	**rancidez**	rancissement
6.2-**66**	risk	**riesgo**	risque
6.2-**67**	risk analysis	**análisis de riesgos**	analyse des risques
6.2-**68**	risk assessment	**evaluación del riesgo**	appréciation des risques
6.2-**69**	risk characterization	**caracterización del riesgo**	caractérisation des risques
6.2-**70**	risk communication	**información sobre riesgos**	communication à propos des risques
6.2-**71**	risk management	**gestión del riesgo**	gestion des risques
6.2-**72**	self-inspection	**inspección**	autocontrôle
6.2-**73**	shrinkage *withering* ○	**marchitamiento**	flétrissure
6.2-**74**	slime	**capa pegajosa** capa viscosa	couche poisseuse *couche visqueuse*
6.2-**75**	smooth	**alisada**	lisse *conforme*
6.2-**76**	soil	**suciedad**	souillure
6.2-**77**	splash area	**área de goteo** área de salpicado	zone d'éclaboussures
6.2-**78**	spoilage *deterioration*	**deterioro** alteración	altération *détérioration* ○
6.2-**79**	storage disease	**enfermedad de almacenamiento**	maladie d'entreposage
6.2-**80**	storage disorder	**alteración durante el almacenamiento**	accident d'entreposage
6.2-**81**	target level	**límite de aceptación**	niveau cible
6.2-**82**	traceability	**trazabilidad**	traçabilité
6.2-**83**	validation	**validación**	validation
6.2-**84**	verification	**verificación**	vérification *contrôle*
6.2-**85**	weight loss	**pérdida de peso**	perte de poids *perte de masse* ○
6.2-**86**	withering ○ *shrinkage*	**marchitamiento**	flétrissure

ENGLISH	ESPAÑOL	FRANÇAIS	
SECTION 6.3 *Refrigeration and the* *food industry* **SECTION 6.3.1** *Products of plant origin*	**SUBCAPÍTULO 6.3** *Tratamiento por el frío en las* *industrias agroalimentarias* **SUBCAPÍTULO 6.3.1** *Productos de origen vegetal*	**SOUS-CHAPITRE 6.3** *Traitement par le froid dans* *l'industrie agroalimentaire* **SOUS-CHAPITRE 6.3.1** *Produits végétaux*	
abnormal external moisture	**agua superficial**	humidité extérieure anormale *embuage* ◐	6.3.1-**1**
accelerated ripening	**maduración acelerada**	maturation accélérée	6.3.1-**2**
acclimatized	**aclimatado**	ressuyé	6.3.1-**3**
alcohol injury	**daño por fermentación alcohólica**	fermentation alcoolique (des fruits)	6.3.1-**4**
artificial atmosphere generator	**generador de atmósfera artificial**	générateur d'atmosphère	6.3.1-**5**
bitter pit	**picado amargo**	taches amères *bitter pit* ◐	6.3.1-**6**
brown core *brown heart* ◐	**corazón pardo**	cœur brun	6.3.1-**7**
brown heart ◐ *brown core*	**corazón pardo**	cœur brun	6.3.1-**8**
C.A. storage ◐ *controlled-atmosphere storage* *storage in controlled atmosphere* ◐	**almacenamiento en AC (Atmósfera Controlada)**	entreposage en atmosphère contrôlée (AC) *entreposage AC*	6.3.1-**9**
C.A. storage room ◐ *controlled-atmosphere (storage) room* *gas store* ○	**cámara de atmósfera controlada**	chambre à atmosphère contrôlée	6.3.1-**10**
chilling damage	**daño por frío**	altération due à la réfrigération	6.3.1-**11**
climacteric rise	**crisis climatérica**	crise climactérique	6.3.1-**12**
controlled-atmosphere storage *C.A. storage* ◐ *storage in controlled atmosphere* ◐	**almacenamiento en AC (Atmósfera Controlada)**	entreposage en atmosphère contrôlée (AC) *entreposage AC*	6.3.1-**13**
controlled-atmosphere (storage) room *C.A. storage room* ◐ *gas store* ○	**cámara de atmósfera controlada**	chambre à atmosphère contrôlée	6.3.1-**14**
core flush	**variante de corazón pardo**	cœur rosé	6.3.1-**15**
cultivar *variety* ◐	**cultivar**	cultivar *variété*	6.3.1-**16**
damage	**daño**	dommage *dégât* ◐	6.3.1-**17**
degree of maturity	**grado de madurez** estado de madurez	degré de maturité	6.3.1-**18**
degreening	**desverdización**	déverdissage	6.3.1-**19**
exchanger-diffuser	**cambiador-difusor**	échangeur-diffuseur	6.3.1-**20**
fruit packing station	**central frutícola de manipulación de frutas** almacén de manipulación de frutas	station fruitière *station de conditionnement* ◐	6.3.1-**21**
fruit and vegetables in bulk	**frutas y hortalizas a granel**	fruits et légumes en vrac	6.3.1-**22**
fruit and vegetables in layers	**envasado de frutas y hortalizas por capas**	fruits et légumes lités	6.3.1-**23**
gas store ○ *controlled-atmosphere (storage) room* *C.A. storage room* ◐	**cámara de atmósfera controlada**	chambre à atmosphère contrôlée	6.3.1-**24**

	ENGLISH	ESPAÑOL	FRANÇAIS
6.3.1-**25**	grading *size grading*	**calibrado** (subst.) clasificación por tamaño	calibrage
6.3.1-**26**	growth defect	**defecto de crecimiento**	défaut de croissance *défaut de développement* ○
6.3.1-**27**	immature	**no madura**	immature
6.3.1-**28**	infrared CO_2-meter	**analizador de CO_2 por rayos infrarrojos**	analyseur de CO_2 par infra-rouge *détecteur par infra-rouge* ○
6.3.1-**29**	internal breakdown *low-temperature breakdown* ○	**descomposición interna**	brunissement interne *décomposition interne* ○ *dégradation interne* ○
6.3.1-**30**	internal defect	**desorden** daño interno	défaut interne
6.3.1-**31**	low-temperature breakdown ○ *internal breakdown*	**descomposición interna**	brunissement interne *décomposition interne* ○ *dégradation interne* ○
6.3.1-**32**	nitrogen flushing	**barrido con nitrógeno**	balayage à l'azote
6.3.1-**33**	Orsat apparatus	**aparato de Orsat**	appareil d'Orsat
6.3.1-**34**	over-mature (vegetable)	**sobremaduro** senescente	montée (d'un légume)
6.3.1-**35**	over-ripeness (of a fruit)	**sobremadurez** senescencia	blettissement (d'un fruit) *surmaturité* ○
6.3.1-**36**	oxygen pull-down	**reducción de la proporción de oxígeno** reducción del porcentaje de oxígeno	abaissement du taux d'oxygène *réduction de la teneur en oxygène*
6.3.1-**37**	paramagnetic O_2-meter	**analizador paramagnético de oxígeno**	analyseur d'oxygène paramagnétique
6.3.1-**38**	physiological disorder	**desorden fisiológico** fisiopatía	maladie physiologique *trouble physiologique*
6.3.1-**39**	physiological maturity (of a fruit)	**madurez fisológica**	maturité physiologique (d'un fruit)
6.3.1-**40**	refrigeration section	**bloque frigorífico** sección frigorífica	bloc frigorifique
6.3.1-**41**	regenerative scrubber	**descarbonizador regenerable**	absorbeur de CO_2 régénérable
6.3.1-**42**	ripening	**maduración**	maturation
6.3.1-**43**	rot	**pudrición** podredumbre	pourriture
6.3.1-**44**	russeting	**ruginosidad** "russeting"	roussissement *rugosité* ○
6.3.1-**45**	scald	**escaldadura superficial**	échaudure
6.3.1-**46**	scrubber	**absorbedor de anhídrido carbónico** descarbonizador	absorbeur de CO_2
6.3.1-**47**	size grading *grading*	**calibrado** (subst.) clasificación por tamaño	calibrage
6.3.1-**48**	storage in controlled atmosphere ○ *controlled-atmosphere storage* *C.A. storage* ○	**almacenamiento en AC (Atmósfera Controlada)**	entreposage en atmosphère contrôlée (AC) *entreposage AC*
6.3.1-**49**	variety ○ *cultivar*	**cultivar**	cultivar *variété*
6.3.1-**50**	vegetable packing station	**central hortícola**	station légumière

ENGLISH	ESPAÑOL	FRANÇAIS	
SECTION 6.3.2 *Meat*	**SUBCAPÍTULO 6.3.2** *Productos cárnicos*	**SOUS-CHAPITRE 6.3.2** *Produits carnés*	
abattoir *slaughterhouse* ○	**matadero**	abattoir	6.3.2-**1**
ageing (of meat) *maturation*	**maduración (de las carnes)**	maturation (de la viande)	6.3.2-**2**
boned meat	**carne deshuesada**	viande désossée	6.3.2-**3**
bone taint	**alteración en profundidad**	altération en profondeur	6.3.2-**4**
brining	**salazón en salmuera**	saumurage	6.3.2-**5**
carcass	**canal**	carcasse	6.3.2-**6**
(carcass) chilling-process	**preenfriamiento de canales** oreo refrigerado	réfrigération primaire (des carcasses) *ressuage réfrigéré* ○	6.3.2-**7**
cold shortening	**contracción muscular por frío**	contraction par le froid *raccourcissement dû au froid* ○	6.3.2-**8**
curing cellar	**cámara de salazón** cámara de curado	chambre de salaison	6.3.2-**9**
eviscerated poultry	**aves evisceradas**	volaille éviscérée	6.3.2-**10**
gutted poultry	**aves destripadas**	volaille effilée	6.3.2-**11**
hot deboning	**deshuesado en caliente**	désossage à chaud	6.3.2-**12**
maturation *ageing (of meat)*	**maduración (de las carnes)**	maturation (de la viande)	6.3.2-**13**
(meat) ageing room ○	**cámara de maduración (de carnes)**	chambre de maturation (de la viande)	6.3.2-**14**
(meat) cutting room	**sala de troceado (de carnes)** sala de despiece	salle de découpe (de viande)	6.3.2-**15**
meat holding room	**cámara de conservación (de carnes)**	chambre de conservation (de viande) *resserre à viande* ○	6.3.2-**16**
offal	**despojos**	abats	6.3.2-**17**
quarter (of meat)	**cuarto (de res)** cuarto (de carne)	quartier (de viande)	6.3.2-**18**
sausage drying room	**secadero de embutidos**	séchoir à saucisson	6.3.2-**19**
shock-chilling	**enfriamiento de choque**	réfrigération-choc	6.3.2-**20**
slaughterhouse ○ *abattoir*	**matadero**	abattoir	6.3.2-**21**
smoking	**ahumado**	fumaison	6.3.2-**22**
spin chiller	**enfriador de tambor-agitador**	refroidisseur à tambour-agitateur	6.3.2-**23**
thaw rigor	**contracción durante la descongelación** rigidez durante la descongelación	rigor à la décongélation *rigidité à la décongélation* ○	6.3.2-**24**
tripe	**mondongo**	tripes	6.3.2-**25**

ENGLISH	ESPAÑOL	FRANÇAIS	
SECTION 6.3.3 *Dairy products*	**SUBCAPÍTULO 6.3.3** *Productos lácteos*	**SOUS-CHAPITRE 6.3.3** *Produits laitiers*	
bulk collection of milk	**recogida de leche a granel**	ramassage du lait en vrac	6.3.3-**1**
bulk milk cooler (direct-expansion type)	**enfriador de leche a granel por expansión directa**	refroidisseur de lait en vrac à détente directe	6.3.3-**2**

	ENGLISH	ESPAÑOL	FRANÇAIS
6.3.3-**3**	bulk milk cooler (iced-water type)	**enfriador de leche a granel por agua helada**	refroidisseur de lait en vrac à eau glacée
6.3.3-**4**	butter factory	**fábrica de mantequilla**	beurrerie
6.3.3-**5**	butter making	**fabricación de mantequilla**	beurrerie
6.3.3-**6**	can milk cooler (USA) *churn milk cooler (UK)*	**enfriador de leche en bidones** enfriador de leche en cántaros	refroidisseur de lait en bidons *refroidisseur de lait en pots* ◐
6.3.3-**7**	can sprinkling cooler *cascade milk cooler* *irrigation cooler*	**enfriador de leche por aspersión**	refroidisseur de lait à aspersion
6.3.3-**8**	cascade milk cooler *can sprinkling cooler* *irrigation cooler*	**enfriador de leche por aspersión**	refroidisseur de lait à aspersion
6.3.3-**9**	central milk plant	**planta lechera**	centrale laitière
6.3.3-**10**	cheese curing *cheese ripening*	**maduración de quesos**	affinage du fromage
6.3.3-**11**	cheese drying	**secado del queso**	hâlage du fromage
6.3.3-**12**	cheese drying room	**secadero de quesos**	hâloir à fromage
6.3.3-**13**	cheese factory	**quesería** fábrica de quesos	fromagerie
6.3.3-**14**	cheese making	**fabricación de queso**	fromagerie
6.3.3-**15**	cheese ripening *cheese curing*	**maduración de quesos**	affinage du fromage
6.3.3-**16**	churn immersion cooler *immersion milk cooler*	**enfriador de leche por inmersión**	refroidisseur de bidons de lait à immersion
6.3.3-**17**	churn milk cooler (UK) *can milk cooler (USA)*	**enfriador de leche en bidones** enfriador de leche en cántaros	refroidisseur de lait en bidons *refroidisseur de lait en pots* ◐
6.3.3-**18**	churning	**batido (operación)**	barattage
6.3.3-**19**	city milk plant	**planta de higienización de leche**	centrale laitière
6.3.3-**20**	clotting *coagulation*	**cuajadura** coagulación	caillage *coagulation*
6.3.3-**21**	coagulation *clotting*	**cuajadura** coagulación	caillage *coagulation*
6.3.3-**22**	curd	**cuajada**	caillé
6.3.3-**23**	dairy factory *milk plant* *dairy plant* *milk factory*	**central lechera**	laiterie (usine) *usine laitière*
6.3.3-**24**	dairy plant *milk plant* *dairy factory* *milk factory*	**central lechera**	laiterie (usine) *usine laitière*
6.3.3-**25**	dairy shop	**lechería**	crémerie *laiterie (boutique)* ◐
6.3.3-**26**	immersion milk cooler *churn immersion cooler*	**enfriador de leche por inmersión**	refroidisseur de bidons de lait à immersion
6.3.3-**27**	in-can immersion cooler *plunger-type milk cooler*	**enfriador de leche con evaporador sumergido amovible**	refroidisseur de lait à plongeur
6.3.3-**28**	irrigation cooler *can sprinkling cooler* *cascade milk cooler*	**enfriador de leche por aspersión**	refroidisseur de lait à aspersion

ENGLISH	ESPAÑOL	FRANÇAIS	
milk collection centre	**centro de recogida de leche**	centre de ramassage de lait *centre de collecte de lait*	6.3.3-**29**
milk collection in cans (USA) *milk collection in churns (UK)*	**recogida de leche en cántaros**	ramassage du lait en bidons *ramassage du lait en pots* ○	6.3.3-**30**
milk collection in churns (UK) *milk collection in cans (USA)*	**recogida de leche en cántaros**	ramassage du lait en bidons *ramassage du lait en pots* ○	6.3.3-**31**
milk cooler	**enfriador de leche**	refroidisseur de lait	6.3.3-**32**
milk cooling on the farm	**enfriamiento de la leche en la granja**	refroidissement du lait à la ferme	6.3.3-**33**
milk factory *milk plant* *dairy factory* *dairy plant*	**central lechera**	laiterie (usine) *usine laitière*	6.3.3-**34**
milk plant *dairy factory* *dairy plant* *milk factory*	**central lechera**	laiterie (usine) *usine laitière*	6.3.3-**35**
plunger-type milk cooler *in-can immersion cooler*	**enfriador de leche con evaporador sumergido amovible**	refroidisseur de lait à plongeur	6.3.3-**36**
refrigerated farm tank	**tanque enfriador de granja**	bac refroidisseur de ferme *cuve de refroidissement à la ferme* ○	6.3.3-**37**
sparge-ring-type milk cooler ○	**enfriador de leche con anillo de aspersión** enfriador de leche con collar de aspersión	refroidisseur de lait à collier d'aspersion	6.3.3-**38**
turbine milk cooler	**enfriador de leche tipo turbina**	refroidisseur de lait à tourniquet	6.3.3-**39**
uperization ○	**uperización**	upérisation	6.3.3-**40**

SECTION 6.3.4 *Ice cream*	SUBCAPÍTULO 6.3.4 *Helados*	SOUS-CHAPITRE 6.3.4 *Crèmes glacées*	
batch-type (ice cream) freezer	**congelador discontinuo (de helados)**	congélateur discontinu (de crème glacée) *turbine à crème glacée* ○	6.3.4-**1**
conservator ○ *ice cream cabinet*	**conservador (de helados)** mueble conservador (de helados)	conservateur de crème glacée	6.3.4-**2**
contact hardening	**endurecimiento por contacto**	durcissement par contact	6.3.4-**3**
continuous (ice cream) freezer	**congelador continuo (de helados)**	congélateur continu (de crème glacée)	6.3.4-**4**
hardening	**endurecimiento (de helados)**	durcissement	6.3.4-**5**
ice cream	**helado**	crème glacée *glace* ○	6.3.4-**6**
ice cream brick	**helado al corte**	pavé de glace	6.3.4-**7**
ice cream cabinet *conservator* ○	**conservador (de helados)** mueble conservador (de helados)	conservateur de crème glacée	6.3.4-**8**
ice cream freezer	**congelador de helados**	congélateur de crème glacée	6.3.4-**9**
(ice cream) hardening cabinet	**mueble de endurecimiento (de helados)**	meuble de durcissement (de crème glacée)	6.3.4-**10**
(ice cream) hardening room	**cámara de endurecimiento de helados**	chambre de durcissement (de crème glacée)	6.3.4-**11**
ice cream manufacturer	**fabricante de helados**	glacier	6.3.4-**12**
(ice cream) mix	**mezcla para helados**	mélange (pour crème glacée) *mix* ○	6.3.4-**13**

	ENGLISH	ESPAÑOL	FRANÇAIS
6.3.4-**14**	ice lolly	**polo**	bâtonnet glacé
6.3.4-**15**	ice milk	**helado de leche**	glace au lait
6.3.4-**16**	over-run	**expansión debida al aire (helados)** aumento de volumen por inclusión de aire	foisonnement
6.3.4-**17**	rapid hardener	**instalación de endurecimiento rápido** endurecedor rápido (de helados)	installation de durcissement rapide
6.3.4-**18**	sherbet (USA) *sorbet* *water ice* ○	**sorbete**	sorbet
6.3.4-**19**	soft ice (cream)	**helado blando (sin endurecer)** helado italiano	glace "à l'italienne" *glace molle* ○
6.3.4-**20**	sorbet *sherbet (USA)* *water ice* ○	**sorbete**	sorbet
6.3.4-**21**	tunnel hardening	**endurecimiento en túnel**	durcissement en tunnel
6.3.4-**22**	water ice ○ *sorbet* *sherbet (USA)*	**sorbete**	sorbet

	SECTION 6.3.5 *Fish and seafood*	SUBCAPÍTULO 6.3.5 *Productos de la pesca*	SOUS-CHAPITRE 6.3.5 *Produits de la pêche*
6.3.5-**1**	autolysis (of fish)	**autolisis**	autolyse
6.3.5-**2**	bilgy fish ○ *stinker* ○	**pescado con olor a fango**	poisson fangeux ○
6.3.5-**3**	chilled sea water (CSW)	**salmuera fría**	eau de mer fraîche
6.3.5-**4**	filleting (of fish)	**fileteado (del pescado)**	filetage (du poisson)
6.3.5-**5**	icing	**envasado con hielo**	glaçage
6.3.5-**6**	refrigerated sea water (RSW)	**salmuera refrigerada**	eau de mer réfrigérée
6.3.5-**7**	rust	**roya** herrumbre	rouille
6.3.5-**8**	stinker ○ *bilgy fish* ○	**pescado con olor a fango**	poisson fangeux ○

	SECTION 6.3.6 *Beverages*	SUBCAPÍTULO 6.3.6 *Bebidas*	SOUS-CHAPITRE 6.3.6 *Boissons*
6.3.6-**1**	beer cooler	**enfriador de cerveza**	refroidisseur de bière
6.3.6-**2**	bottom fermentation *low fermentation* ○	**fermentación baja**	fermentation basse
6.3.6-**3**	carbonator	**carbonatador**	saturateur (en CO_2)
6.3.6-**4**	chill proofing *cold stabilization* ○	**estabilización por el frío**	stabilisation par le froid
6.3.6-**5**	cold stabilization ○ *chill proofing*	**estabilización por el frío**	stabilisation par le froid
6.3.6-**6**	fermenting cellar	**bodega de fermentación**	salle de fermentation

ENGLISH	ESPAÑOL	FRANÇAIS	
high fermentation ○ *top fermentation*	**fermentación alta**	fermentation haute	6.3.6-**7**
low fermentation ○ *bottom fermentation*	**fermentación baja**	fermentation basse	6.3.6-**8**
stock cellar	**bodega de almacenamiento** bodega de guarda	cave de garde	6.3.6-**9**
top fermentation *high fermentation* ○	**fermentación alta**	fermentation haute	6.3.6-**10**
wine factory ○ *winery*	**bodega de vinificación**	cave de vinification	6.3.6-**11**
winery *wine factory* ○	**bodega de vinificación**	cave de vinification	6.3.6-**12**
wort cooler	**enfriador de mosto (de cerveza)**	refroidisseur de moût	6.3.6-**13**

SECTION 6.3.7 *Other products* *(egg products, sweets, etc.)*	SUBCAPÍTULO 6.3.7 *Otros productos* *(de huevo, confitería, etc.)*	SOUS-CHAPITRE 6.3.7 *Autres produits* *(ovoproduits, confiserie, etc.)*	
bakery (refrigerated) slab	**mesa refrigerada para pastelería**	table froide *tour de pâtisserie* ○	6.3.7-**1**
bottomer slab	**placa de endurecimiento de fondos**	plaque de durcissement des fonds	6.3.7-**2**
egg-breaking plant (USA) ○ *egg-shelling plant*	**equipo para rotura de huevos**	casserie d'œufs	6.3.7-**3**
egg candling	**ovoscopia**	mirage des œufs	6.3.7-**4**
egg-shelling plant *egg-breaking plant (USA)* ○	**equipo para rotura de huevos**	casserie d'œufs	6.3.7-**5**
sponge dough	**masa "semidirecta"** masa semipreparada	pâte pain-levure *pâte "semi-directe"* ○	6.3.7-**6**
staling	**correosidad**	rassissement	6.3.7-**7**
straight dough	**masa "directa"** masa terminada	pâte directe	6.3.7-**8**
winterisation	**enturbiamento (intencionado) de los aceites por el frío** "winterización"	frigélisation	6.3.7-**9**

SECTION 6.4 *Other treatments used in the* *food industry*	SUBCAPÍTULO 6.4 *Otros tratamientos en las industrias* *agroalimentarias*	SOUS-CHAPITRE 6.4 *Autres traitements dans l'industrie* *agroalimentaire*	
adjunct to refrigeration *supplement to refrigeration* ○	**coadyuvante del frío**	adjuvant du froid	6.4-**1**
alginate coating	**recubrimiento con alginatos**	revêtement gélifiant	6.4-**2**
anti-oxidant	**antioxidante**	anti-oxydant *anti-oxygène* ○	6.4-**3**
antibiotic	**antibiótico**	antibiotique	6.4-**4**
appertization	**apertización**	appertisation	6.4-**5**

	ENGLISH	ESPAÑOL	FRANÇAIS
6.4-**6**	bactericide	**bactericida** (subst.)	bactéricide
6.4-**7**	bacteriostatic	**bacteriostático**	bactériostatique
6.4-**8**	blanching	**escaldado**	blanchiment
6.4-**9**	camouflage of goods	**camuflage**	fardage
6.4-**10**	canned food (USA) *tinned food* ○	**alimento en conserva** alimento enlatado	conserve appertisée
6.4-**11**	food additive	**aditivo alimentario**	additif alimentaire
6.4-**12**	fumigation	**fumigación**	fumigation
6.4-**13**	fungicide	**fungicida** (subst.)	fongicide
6.4-**14**	gel *jelly* ○	**gel** jalea	gel *gelée* ○
6.4-**15**	gelation	**gelificación**	gélification
6.4-**16**	ionizing radiation	**radiaciones ionizantes**	rayonnements ionisants
6.4-**17**	irradiation	**irradiación**	irradiation
6.4-**18**	jelly ○ *gel*	**gel** jalea	gel *gelée* ○
6.4-**19**	odour control	**lucha contra los olores**	lutte contre les odeurs
6.4-**20**	oiling	**aceitado** (subst.)	huilage
6.4-**21**	pasteurization	**pasterización**	pasteurisation
6.4-**22**	preheating	**precalentamiento**	préchauffage
6.4-**23**	preservative	**agente conservador** aditivo conservador	agent conservateur *conservateur*
6.4-**24**	radiation pasteurization *radiopasteurization* ○ *radurization* ○	**radiopasterización**	radiopasteurisation
6.4-**25**	radiation sterilization *radioappertization* ○ *radiosterilization* ○	**radioesterilización**	radiostérilisation *radappertisation* ○ *radioappertisation* ○
6.4-**26**	radicidation	**radicidación**	radicidation
6.4-**27**	radioappertization ○ *radiation sterilization* *radiosterilization* ○	**radioesterilización**	radiostérilisation *radappertisation* ○ *radioappertisation* ○
6.4-**28**	radiopasteurization ○ *radiation pasteurization* *radurization* ○	**radiopasterización**	radiopasteurisation
6.4-**29**	radiosterilization ○ *radiation sterilization* *radioappertization* ○	**radioesterilización**	radiostérilisation *radappertisation* ○ *radioappertisation* ○
6.4-**30**	radurization ○ *radiation pasteurization* *radiopasteurization* ○	**radiopasterización**	radiopasteurisation
6.4-**31**	scalding	**escaldado** (subst.)	échaudage
6.4-**32**	semi-preserve	**semiconserva**	semi-conserve
6.4-**33**	steaming	**escaldado al vapor**	ébouillantage
6.4-**34**	sterilization	**esterilización**	stérilisation
6.4-**35**	supplement to refrigeration ○ *adjunct to refrigeration*	**coadyuvante del frío**	adjuvant du froid

ENGLISH	ESPAÑOL	FRANÇAIS	
tinned food o *canned food (USA)*	**alimento en conserva** alimento enlatado	conserve appertisée	6.4-**36**
warming room	**sala de acondicionamiento térmico** cámara de recalentamiento	salle de réchauffage	6.4-**37**
wax coating	**recubrimiento céreo** recubrimiento de parafina	revêtement cireux *revêtement paraffineux* o	6.4-**38**

SECTION 6.5 *Packaging*	SUBCAPÍTULO 6.5 *Acondicionado y embalaje*	SOUS-CHAPITRE 6.5 *Conditionnement et emballage*	
active packaging	**envasado activo**	emballage actif *emballage en atmosphère modifiée* o	6.5-**1**
batch *lot*	**lote**	lot	6.5-**2**
lot *batch*	**lote**	lot	6.5-**3**
marking	**etiquetado**	marquage	6.5-**4**
modified atmosphere packaging (MAP)	**envasado en atmosfera modificada (EAM)**	emballage sous atmosphère modifiée	6.5-**5**
packaging *packing*	**embalaje** envasado	conditionnement *emballage*	6.5-**6**
packing *packaging*	**embalaje** envasado	conditionnement *emballage*	6.5-**7**
packing station	**almacen de envasado**	centre de conditionnement *station de conditionnement* *plate-forme de conditionnement* o *station d'emballage* o	6.5-**8**
prepackaging o *prepacking*	**preembalaje**	préemballage	6.5-**9**
prepacking *prepackaging* o	**preembalaje**	préemballage	6.5-**10**
sorting	**clasificación**	triage	6.5-**11**
transport packaging	**empaquetado para transporte**	suremballage	6.5-**12**
ungrouping	**desagrupado**	dégroupage	6.5-**13**
vacuum packing	**envasado al vacio**	conditionnement sous vide *emballage sous vide* o	6.5-**14**

Capítulo 7.

ACONDICIONAMIENTO DE AIRE

- ◑ término aceptado
- ○ término obsoleto

ENGLISH	ESPAÑOL	FRANÇAIS	
SECTION 7.1 *Air conditioning:* *general background*	**SUBCAPÍTULO 7.1** *Generalidades sobre* *acondicionamiento de aire*	**SOUS-CHAPITRE 7.1** *Généralités sur le* *conditionnement d'air*	
acclimation (USA) *acclimatization*	**aclimatación**	acclimatation	7.1-**1**
acclimatization *acclimation (USA)*	**aclimatación**	acclimatation	7.1-**2**
air conditioning (AC)	**acondicionamiento de aire**	conditionnement d'air	7.1-**3**
air-conditioning installation *air-conditioning plant*	**instalación de acondicionamiento de aire** instalación de climatización	installation de conditionnement d'air	7.1-**4**
air-conditioning plant *air-conditioning installation*	**instalación de acondicionamiento de aire** instalación de climatización	installation de conditionnement d'air	7.1-**5**
air-conditioning process *air-conditioning system*	**procedimiento de acondicionamiento de aire**	procédé de conditionnement d'air *système de conditionnement d'air*	7.1-**6**
air-conditioning system *air-conditioning process*	**procedimiento de acondicionamiento de aire**	procédé de conditionnement d'air *système de conditionnement d'air*	7.1-**7**
air handling *air treatment*	**tratamiento del aire**	traitement de l'air	7.1-**8**
air treatment *air handling*	**tratamiento del aire**	traitement de l'air	7.1-**9**
all-year air conditioning *year-round air conditioning*	**acondicionamiento del aire para invierno y verano** acondicionamiento del aire para toda estación	conditionnement d'air toutes saisons *conditionnement d'air été-hiver* ○	7.1-**10**
ambient air	**aire ambiente**	air ambiant	7.1-**11**
apparatus dew point	**punto de rocío equivalente**	point de rosée équivalent	7.1-**12**
atmosphere of reference ○	**atmósfera de referencia**	atmosphère de référence ○	7.1-**13**
chiller	**máquina enfriadora**	groupe refroidisseur d'eau	7.1-**14**
climatic engineering ○ *environmental engineering* ○	**ingeniería de climatización**	génie climatique	7.1-**15**
comfort air conditioning	**acondicionamiento para (el) confort** climatización	climatisation *conditionnement d'air de confort* ○	7.1-**16**
comfort cooling	**enfriamiento para confort**	rafraîchissement pour le confort	7.1-**17**
comfort index	**índice de confort**	indice de confort	7.1-**18**
conditioned air	**aire acondicionado**	air conditionné	7.1-**19**
corrected effective temperature	**temperatura efectiva corregida**	température effective corrigée	7.1-**20**
degree-day *kelvin-day* ○	**grados-día** kelvin-día	degré-jour	7.1-**21**
dehumidifying effect	**capacidad de deshumidificación**	puissance de déshumidification *puissance frigorifique latente (d'un refroidisseur d'air)* ○	7.1-**22**
design conditions	**condiciones de proyecto**	conditions contractuelles	7.1-**23**
effective temperature	**temperatura efectiva**	température effective	7.1-**24**
environment cooling ○	**enfriamiento de ambientes** enfriamiento del medio ambiente	refroidissement d'une ambiance	7.1-**25**
environmental conditions	**condiciones ambientales**	conditions d'ambiance	7.1-**26**

ENGLISH	ESPAÑOL	FRANÇAIS
7.1-27 environmental engineering ◐ *climatic engineering* ○	**ingeniería de climatización**	génie climatique
7.1-28 equivalent temperature	**temperatura equivalente**	température équivalente *température résultante sèche*
7.1-29 eupatheoscope	**eupateóscopo**	eupathéoscope
7.1-30 freshness index	**índice de frescor**	indice de fraîcheur
7.1-31 frigorimeter	**frigorímetro**	frigorimètre
7.1-32 globe thermometer	**térmometro de globo**	thermomètre globe
7.1-33 greenhouse effect	**efecto de invernadero**	effet de serre
7.1-34 humidifying effect	**capacidad de humidificación**	puissance d'humidification
7.1-35 indoor climate	**condiciones ambientales interiores** ambiente interior	climat intérieur *climat confiné*
7.1-36 indoor conditions	**condiciones interiores**	conditions intérieures
7.1-37 industrial air conditioning *process air conditioning* ◐	**acondicionamiento de aire industrial**	conditionnement d'air industriel
7.1-38 insolation	**insolación**	insolation
7.1-39 katathermometer	**catatermómetro**	catathermomètre
7.1-40 kelvin-day ◐ *degree-day*	**grados-día** kelvin-día	degré-jour
7.1-41 latent cooling capacity	**potencia frigorífica latente**	puissance frigorifique latente
7.1-42 leakage airflow	**fuga de aire**	fuite d'air
7.1-43 low-grade heat source	**fuente de calor de poca intensidad**	source de chaleur à basse température
7.1-44 marine air conditioning	**acondicionamiento del aire para buques** climatización de buques	conditionnement d'air à bord des navires
7.1-45 microclimate	**microclima**	microclimat
7.1-46 multizone	**multizona**	multizone
7.1-47 net cooling capacity	**potencia frigorífica neta**	puissance frigorifique nette
7.1-48 net total cooling capacity	**potencia frigorífica total neta**	puissance frigorifique totale nette
7.1-49 occupied zone	**zona ocupada**	zone occupée
7.1-50 outdoor conditions	**condiciones exteriores**	conditions extérieures
7.1-51 process air conditioning ◐ *industrial air conditioning*	**acondicionamiento de aire industrial**	conditionnement d'air industriel
7.1-52 residential air conditioning	**acondicionamiento de aire para viviendas**	conditionnement d'air résidentiel
7.1-53 resulting temperature	**temperatura resultante**	température résultante
7.1-54 sensible cooling capacity *sensible (dry) air-cooling capacity* *sensible cooling effect* ○	**potencia frigorífica sensible** potencia frigorífica sensible del aire potencia frigorífica sensible (de un enfriador de aire)	puissance frigorifique sensible
7.1-55 sensible (dry) air-cooling capacity *sensible cooling capacity* *sensible cooling effect* ○	**potencia frigorífica sensible** potencia frigorífica sensible del aire potencia frigorífica sensible (de un enfriador de aire)	puissance frigorifique sensible
7.1-56 sensible cooling effect ○ *sensible cooling capacity* *sensible (dry) air-cooling capacity*	**potencia frigorífica sensible** potencia frigorífica sensible del aire potencia frigorífica sensible (de un enfriador de aire)	puissance frigorifique sensible

ENGLISH	ESPAÑOL	FRANÇAIS	
sensible heat ratio	**factor de calor sensible (de un enfriador de aire)**	coefficient de chaleur sensible (d'un refroidisseur d'air)	7.1-**57**
solar heat gain	**aporte de calor por insolación**	apport de chaleur par insolation	7.1-**58**
standard air	**aire normalizado**	air normal	7.1-**59**
standard atmosphere of reference	**atmósfera normal de referencia**	atmosphère normale de référence	7.1-**60**
total cooling capacity *total cooling effect* ○	**potencia frigorífica total** potencia frigorífica total (de un enfriador de aire)	puissance frigorifique totale	7.1-**61**
total cooling effect ○ *total cooling capacity*	**potencia frigorífica total** potencia frigorífica total (de un enfriador de aire)	puissance frigorifique totale	7.1-**62**
treated air	**aire tratado**	air traité	7.1-**63**
year-round air conditioning *all-year air conditioning*	**acondicionamiento del aire para invierno y verano** acondicionamiento del aire para toda estación	conditionnement d'air toutes saisons *conditionnement d'air été-hiver* ◐	7.1-**64**
zoning	**zonificación**	zonage	7.1-**65**

SECTION 7.2 *Specific definitions* **SECTION 7.2.1** *Air-conditioning production*	**SUBCAPÍTULO 7.2** *Conocimientos específicos* **SUBCAPÍTULO 7.2.1** *Producción del acondicionamiento de aire*	**SOUS-CHAPITRE 7.2** *Notions spécifiques* **SOUS-CHAPITRE 7.2.1** *Production d'air conditionné*	
air cooler *cooling unit* *unit cooler*	**enfriador de aire** frigorífero	refroidisseur d'air *frigorifère* ◐	7.2.1-**1**
bypassed indoor airflow	**caudal interior recirculado** caudal interior bypasado	air intérieur recyclé	7.2.1-**2**
bypassed outdoor airflow	**caudal exterior recirculado** caudal exterior bypasado	air extérieur recyclé	7.2.1-**3**
cooling unit *air cooler* *unit cooler*	**enfriador de aire** frigorífero	refroidisseur d'air *frigorifère* ◐	7.2.1-**4**
desiccant cooling *desiccant evaporative cooling*	**sistema de enfriamiento evaporativo con desecación**	système déshydratant *système déshydratant à évaporation*	7.2.1-**5**
desiccant evaporative cooling *desiccant cooling*	**sistema de enfriamiento evaporativo con desecación**	système déshydratant *système déshydratant à évaporation*	7.2.1-**6**
double-spacing finned cooler ◐ *two-way finned cooler*	**enfriador de doble paso de aletas**	refroidisseur à double écartement d'ailettes	7.2.1-**7**
draught tower	**enfriamiento por ventilación con efecto chimenea**	tour à vent	7.2.1-**8**
dry cooling coil *dry-surface coil*	**batería seca**	batterie sèche	7.2.1-**9**
dry-surface coil *dry cooling coil*	**batería seca**	batterie sèche	7.2.1-**10**

	ENGLISH	ESPAÑOL	FRANÇAIS
7.2.1-**11**	dry-type air cooler	**frigorígeno seco** enfriador seco de aire	refroidisseur d'air du type sec *frigorifère sec* ⊙
7.2.1-**12**	forced-circulation air cooler *forced-convection air cooler* ⊙ *forced-draught air cooler* ⊙	**enfriador de aire por convección forzada**	refroidisseur d'air à convection forcée *aérofrigorifère* ⊙
7.2.1-**13**	forced-convection air cooler ⊙ *forced-circulation air cooler* *forced-draught air cooler* ⊙	**enfriador de aire por convección forzada**	refroidisseur d'air à convection forcée *aérofrigorifère* ⊙
7.2.1-**14**	forced-convection air heater ○ *unit heater*	**aerotermo**	aérotherme *réchauffeur d'air à convection forcée*
7.2.1-**15**	forced-draught air cooler ⊙ *forced-circulation air cooler* *forced-convection air cooler* ⊙	**enfriador de aire por convección forzada**	refroidisseur d'air à convection forcée *aérofrigorifère* ⊙
7.2.1-**16**	gas-fired heater	**calentamiento directo por gas**	dispositif de chauffage au gaz
7.2.1-**17**	heater battery	**batería de calefacción**	batterie de chauffe
7.2.1-**18**	heater coil *heating resistance* ⊙	**resistencia calefactora**	résistance chauffante
7.2.1-**19**	heating coil	**serpentín de calefacción**	serpentin de chauffage *tube chauffant*
7.2.1-**20**	heating resistance ⊙ *heater coil*	**resistencia calefactora**	résistance chauffante
7.2.1-**21**	indoor discharge airflow	**aire de impulsión**	air intérieur soufflé
7.2.1-**22**	indoor heat exchanger	**cambiador de calor interior**	échangeur thermique intérieur
7.2.1-**23**	indoor intake airflow	**aire de retorno**	air intérieur repris
7.2.1-**24**	natural-convection air cooler	**enfriador de aire por convección natural**	refroidisseur d'air à convection naturelle
7.2.1-**25**	non-ducted indoor air-conditioning equipment	**equipo de acondicionamiento de aire condescarga directa**	climatiseur intérieur non raccordé
7.2.1-**26**	outdoor-discharge airflow	**aire descargado al exterior**	air extérieur refoulé
7.2.1-**27**	outdoor heat exchanger	**cambiador de calor exterior**	échangeur thermique extérieur
7.2.1-**28**	outdoor intake airflow	**aire exterior**	air extérieur aspiré
7.2.1-**29**	panel cooler	**panel de enfriamiento**	panneau refroidisseur
7.2.1-**30**	preheating	**precalentamiento**	préchauffage
7.2.1-**31**	radiant cooling system	**sistema de enfriamiento por paneles**	système de refroidissement par panneaux
7.2.1-**32**	reheating	**recalentamiento (del aire)**	réchauffage
7.2.1-**33**	self-contained air cooler *self-contained cooling unit*	**frigorígeno con grupo incorporado** enfriador de aire con grupo incorporado	refroidisseur d'air à groupe incorporé
7.2.1-**34**	self-contained cooling unit *self-contained air cooler*	**frigorígeno con grupo incorporado** enfriador de aire con grupo incorporado	refroidisseur d'air à groupe incorporé
7.2.1-**35**	sensible-heat air cooler	**frigorígeno de calor sensible** enfriador de aire de calor sensible	refroidisseur d'air à chaleur sensible
7.2.1-**36**	spray-type air cooler	**frigorígeno de pulverización**	refroidisseur d'air à pulvérisation *frigorifère à pulvérisation* ⊙
7.2.1-**37**	troffer ⊙ *ventilated light fitting*	**luminaria ventilada**	luminaire ventilé
7.2.1-**38**	two-way finned cooler *double spacing finned cooler* ⊙	**enfriador de doble paso de aletas**	refroidisseur à double écartement d'ailettes

ENGLISH	ESPAÑOL	FRANÇAIS	
unit cooler *air cooler* *cooling unit*	**enfriador de aire** frigorífero	refroidisseur d'air *frigorifère* ○	7.2.1-**39**
unit heater *forced-convection air heater* ○	**aerotermo**	aérotherme *réchauffeur d'air à convection forcée*	7.2.1-**40**
ventilated light fitting *troffer* ○	**luminaria ventilada**	luminaire ventilé	7.2.1-**41**
wet-type air cooler	**frigorígeno húmedo** enfriador húmedo de aire	refroidisseur d'air du type humide *frigorifère humide* ○	7.2.1-**42**

SECTION 7.2.2 *Air circulation and distribution*	SUBCAPÍTULO 7.2.2 *Movimiento y distribución del aire*	SOUS-CHAPITRE 7.2.2 *Circulation et distribution de l'air*	
aeration	**aireación**	aération	7.2.2-**1**
air change *ventilation rate*	**renovación de aire** caudal de renovación de aire	renouvellement d'air	7.2.2-**2**
air circulation	**circulación de aire**	circulation d'air	7.2.2-**3**
air-circulation ratio *rate of air circulation*	**caudal de recirculación** coeficiente de recirculación	coefficient de brassage *taux de brassage*	7.2.2-**4**
air diffusion	**difusión de aire**	diffusion d'air	7.2.2-**5**
air distribution	**distribución de aire**	distribution d'air	7.2.2-**6**
(air) exfiltration	**fuga (de aire)**	exfiltration (d'air)	7.2.2-**7**
air exhaust *air extract*	**evacuación de aire**	évacuation d'air	7.2.2-**8**
air extract *air exhaust*	**evacuación de aire**	évacuation d'air	7.2.2-**9**
(air) infiltration	**infiltración (de aire)**	infiltration (d'air)	7.2.2-**10**
air inlet *air intake*	**toma de aire** entrada de aire	entrée d'air (neuf) *prise d'air (neuf)*	7.2.2-**11**
air intake *air inlet*	**toma de aire** entrada de aire	entrée d'air (neuf) *prise d'air (neuf)*	7.2.2-**12**
air velocity	**velocidad de aire**	vitesse d'air	7.2.2-**13**
airflow resistance	**resistencia a la corriente del aire**	résistance à l'écoulement de l'air	7.2.2-**14**
airing	**aireación**	aération	7.2.2-**15**
attic ventilation	**ventilación del espacio interior de las cerchas**	ventilation des combles	7.2.2-**16**
axial velocity	**velocidad axial**	vitesse axiale	7.2.2-**17**
blow ○ *throw*	**alcance**	portée	7.2.2-**18**
chimney effect *stack effect* ○	**efecto de chimenea**	tirage	7.2.2-**19**
Coanda effect	**efecto Coanda**	effet Coanda	7.2.2-**20**
cooling air	**aire de enfriamiento**	air de refroidissement	7.2.2-**21**
cross ventilation	**ventilación transversal**	ventilation transversale	7.2.2-**22**
dead air pocket ○ *dead zone* ○	**zona muerta**	zone morte	7.2.2-**23**

ENGLISH	ESPAÑOL	FRANÇAIS
7.2.2-24 dead zone ○ *dead air pocket* ●	**zona muerta**	zone morte
7.2.2-25 delivery air *supply air*	**aire impulsado** aire suministrado	air fourni *air soufflé*
7.2.2-26 diffusion area	**área de difusión**	surface balayée *surface ventilée* ●
7.2.2-27 displacement air diffusion	**difusión de aire por desplazamiento**	diffusion d'air par déplacement
7.2.2-28 draft (USA) *draught*	**tiro** corriente de aire	courant d'air *appel d'air*
7.2.2-29 draught *draft (USA)*	**tiro** corriente de aire	courant d'air *appel d'air*
7.2.2-30 drop	**caída**	retombée
7.2.2-31 entrainment ● *induction*	**inducción**	induction
7.2.2-32 entrainment ratio ● *induction ratio*	**relación de inducción**	taux d'induction *coefficient d'induction* ●
7.2.2-33 exhaust air *extracted air* ●	**aire evacuado**	air évacué *air extrait*
7.2.2-34 exhaust airflow	**caudal de aire de extracción**	air rejeté
7.2.2-35 exit air	**aire expulsado**	air rejeté
7.2.2-36 extracted air ● *exhaust air*	**aire evacuado**	air évacué *air extrait*
7.2.2-37 face velocity *frontal velocity* ●	**velocidad frontal**	vitesse frontale
7.2.2-38 forced-air circulation	**ventilación forzada** circulación forzada (del aire)	ventilation forcée *circulation d'air forcée*
7.2.2-39 (fresh) air make-up	**aire de reposición**	air d'appoint
7.2.2-40 fresh air *outdoor air* *replacement air* *outside air* ●	**aire exterior** aire fresco aire de renovación	air hygiénique *air neuf* *air extérieur* ● *air frais* ●
7.2.2-41 frontal velocity ● *face velocity*	**velocidad frontal**	vitesse frontale
7.2.2-42 induction *entrainment* ●	**inducción**	induction
7.2.2-43 induction ratio *entrainment ratio* ●	**relación de inducción**	taux d'induction *coefficient d'induction* ●
7.2.2-44 mixing air diffusion	**difusión de aire por mezcla**	diffusion d'air par mélange
7.2.2-45 natural air circulation	**circulación natural de aire**	circulation d'air naturelle
7.2.2-46 non-isothermal jet	**chorro no isotermo**	jet non isotherme
7.2.2-47 outdoor air *fresh air* *replacement air* *outside air* ●	**aire exterior** aire fresco aire de renovación	air hygiénique *air neuf* *air extérieur* ● *air frais* ●
7.2.2-48 outside air ● *fresh air* *outdoor air* *replacement air*	**aire exterior** aire fresco aire de renovación	air hygiénique *air neuf* *air extérieur* ● *air frais* ●
7.2.2-49 primary air	**aire primario**	air primaire
7.2.2-50 radius of diffusion	**radio de difusión**	rayon de diffusion

ENGLISH	ESPAÑOL	FRANÇAIS	
rate of air circulation *air-circulation ratio*	**caudal de recirculación** coeficiente de recirculación	coefficient de brassage *taux de brassage*	7.2.2-**51**
recirculated air	**aire recirculado**	air recyclé *air recirculé* ⊙	7.2.2-**52**
replacement air ⊙ *fresh air* *outdoor air* *outside air* ⊙	**aire exterior** aire fresco aire de renovación	air hygiénique *air neuf* *air extérieur* ⊙ *air frais* ⊙	7.2.2-**53**
return air	**aire de retorno**	air repris	7.2.2-**54**
secondary air	**aire secundario**	air secondaire	7.2.2-**55**
spread	**amplitud**	étalement	7.2.2-**56**
stack effect ⊙ *chimney effect*	**efecto de chimenea**	tirage	7.2.2-**57**
supply air *delivery air*	**aire impulsado** aire suministrado	air fourni *air soufflé*	7.2.2-**58**
terminal velocity	**velocidad terminal**	vitesse terminale	7.2.2-**59**
throw *blow* ⊙	**alcance**	portée	7.2.2-**60**
ventilation	**ventilación**	ventilation	7.2.2-**61**
ventilation airflow	**caudal de aire de ventilación**	air neuf	7.2.2-**62**
ventilation rate *air change*	**renovación de aire** caudal de renovación de aire	renouvellement d'air	7.2.2-**63**

SECTION 7.2.3 *Air quality*	SUBCAPÍTULO 7.2.3 *Calidad del aire*	SOUS-CHAPITRE 7.2.3 *Qualité de l'air*	
aerosol	**aerosol**	aérosol	7.2.3-**1**
air contaminant	**contaminante del aire** agente de contaminación del aire	agent de contamination de l'air *contaminant de l'air* *polluant de l'air*	7.2.3-**2**
air pollutant	**contaminante en aire**	polluant de l'air	7.2.3-**3**
airborne particles *particulates (USA)*	**partículas en suspensión** partículas en suspensión (en el aire)	particules en suspension (dans l'air)	7.2.3-**4**
biological agent	**agente biológico**	contaminant biologique	7.2.3-**5**
breathing zone	**zona de respiración**	zone respiratoire	7.2.3-**6**
chemical agent	**agente químico**	contaminant chimique	7.2.3-**7**
decipol	**decipol**	décipol	7.2.3-**8**
deodorization *deodorizing* *odour removal* ⊙	**desodorización** desodoración	désodorisation	7.2.3-**9**
deodorizing *deodorization* *odour removal* ⊙	**desodorización** desodoración	désodorisation	7.2.3-**10**
exposure (by inhalation)	**exposición (por inhalación)**	exposition (par inhalation)	7.2.3-**11**
indoor air quality (IAQ)	**calidad del aire interior**	qualité de l'air intérieur	7.2.3-**12**
odour	**olor**	odeur	7.2.3-**13**

	ENGLISH	ESPAÑOL	FRANÇAIS
7.2.3-**14**	odour removal ❍ *deodorization* *deodorazing*	**desodorización** desodoración	désodorisation
7.2.3-**15**	olf	**olf**	olf
7.2.3-**16**	particulates (USA) *airborne particles*	**partículas en suspensión** partículas en suspensión (en el aire)	particules en suspension (dans l'air)
7.2.3-**17**	separation efficiency	**rendimiento de un filtro**	rendement d'un filtre
7.2.3-**18**	smell	**mal olor**	mauvaise odeur
7.2.3-**19**	suspended matter	**partículas en suspensión**	matières en suspension

	SECTION 7.3 *Specific equipment* **SECTION 7.3.1** *Humidity and temperature control*	**SUBCAPÍTULO 7.3** *Equipos específicos* **SUBCAPÍTULO 7.3.1** *Regulación de la humedad y de la temperatura*	**SOUS-CHAPITRE 7.3** *Matériels spécifiques* **SOUS-CHAPITRE 7.3.1** *Régulation de l'hygrométrie et de la température*
7.3.1-**1**	activated alumina	**alúmina activada**	alumine activée
7.3.1-**2**	air washer *scrubber*	**lavador de aire**	laveur d'air
7.3.1-**3**	atomize (to)	**atomizar** pulverizar	atomiser
7.3.1-**4**	capillary air washer *capillary humidifier* *cell-type air washer*	**lavador alveolar de aire** lavador celular de aire	laveur d'air à alvéoles *humidificateur à alvéoles*
7.3.1-**5**	capillary humidifier *capillary air washer* *cell-type air washer*	**lavador alveolar de aire** lavador celular de aire	laveur d'air à alvéoles *humidificateur à alvéoles*
7.3.1-**6**	cell-type air washer *capillary air washer* *capillary humidifier*	**lavador alveolar de aire** lavador celular de aire	laveur d'air à alvéoles *humidificateur à alvéoles*
7.3.1-**7**	dehumidification capacity	**potencia de deshumectación**	pouvoir de déshumidification
7.3.1-**8**	dehumidification efficiency ratio (DER)	**rendimiento en deshumectación**	coefficient d'efficacité de déshumidification
7.3.1-**9**	dehumidification for comfort	**deshumectación para confort**	déshumidification pour confort
7.3.1-**10**	dehumidification for process	**deshumectación para proceso**	déshumidification pour procédé
7.3.1-**11**	dehumidifier	**deshumidificador** (subst.) desecador (subst.)	déshumidificateur
7.3.1-**12**	desiccant	**desecante**	déshydratant
7.3.1-**13**	desiccant contactor	**sección de deshidratación** sección de secado	contacteur du déshydratant
7.3.1-**14**	desiccant wheel	**rueda desecante** rueda de secado rueda de deshidratación	roue déshydratante
7.3.1-**15**	drift	**agua de arrastre en gotas**	eau entraînée *entraînement vésiculaire* ❍
7.3.1-**16**	eliminator	**separador de gotas**	séparateur de gouttelettes *éliminateur de gouttelettes* ❍

ENGLISH	ESPAÑOL	FRANÇAIS	
fan-pad system	**enfriador de aire con relleno húmedo**	refroidisseur d'air à tampon humide *bourrage ventilé*	7.3.1-**17**
humidification rate	**caudal de humectación**	taux d'humidification	7.3.1-**18**
humidifier	**humidificador** (subst.)	humidificateur	7.3.1-**19**
injection (steam) humidifier	**humidificador de inyección de vapor**	humidificateur à injection de vapeur	7.3.1-**20**
liquid desiccant	**desecante líquido**	déshydratant liquide *solution déshydratante*	7.3.1-**21**
liquid desiccant concentration	**concentración de desecante líquido**	concentration en déshydratant liquide	7.3.1-**22**
liquid desiccant transfer to conditioner	**transferencia de desecante liquido al acondicionador**	transfert du déshydratant liquide au conditionneur	7.3.1-**23**
mean radiant temperature	**temperatura radiante media**	température radiante moyenne	7.3.1-**24**
moisture-removal capacity	**potencia de deshumectación**	rendement d'enlèvement de l'humidité	7.3.1-**25**
moisture-removal rate	**caudal de deshumectación**	taux d'enlèvement de l'humidité	7.3.1-**26**
molecular sieve	**criba molecular** tamiz molecular	tamis moléculaire *crible moléculaire* ○	7.3.1-**27**
pad	**medio dispersante**	milieu dispersant *médium dispersant*	7.3.1-**28**
plane radiant temperature	**temperatura radiante plana**	température plane radiante	7.3.1-**29**
process air	**aire tratado**	air à déshumidifier	7.3.1-**30**
regain (of moisture)	**recuperación (de humedad)**	reprise (d'humidité)	7.3.1-**31**
regeneration air	**aire de regeneración**	air de régénération	7.3.1-**32**
regeneration specific heat input	**consumo específico de regeneración**	chaleur de régénération	7.3.1-**33**
regenerator	**regenerador**	régénérateur	7.3.1-**34**
scrubber *air washer*	**lavador de aire**	laveur d'air	7.3.1-**35**
silica aerogel	**aerosilicagel** aerogel de sílice	aérosilicagel	7.3.1-**36**
spinning disc humidifier	**humidificador de disco giratorio**	humidificateur à disque tournant	7.3.1-**37**
spray chamber	**cámara de pulverización**	chambre de pulvérisation	7.3.1-**38**
spray nozzle	**pulverizador** (subst.)	pulvérisateur	7.3.1-**39**
spray-type air washer	**lavador de aire por pulverización**	laveur d'air à pulvérisation	7.3.1-**40**
surface dehumidifier	**deshumidificador de superficie** desecador de superficie	déshumidificateur à action de surface	7.3.1-**41**

SECTION 7.3.2 *Air circulation and distribution: specific equipment*	SUBCAPÍTULO 7.3.2 *Movimiento y distribución del aire: equipos específicos*	SOUS-CHAPITRE 7.3.2 *Circulation et distribution de l'air: équipements spécifiques*	
air diffuser	**difusor de aire**	diffuseur d'air	7.3.2-**1**
air duct *trunking (1)* ○	**conducto de aire** canal de aire	gaine d'air *conduit d'air* ○	7.3.2-**2**
air grille *grille*	**rejilla de aire**	grille à air	7.3.2-**3**
air-heating fan-coil unit	**ventiloconvector para calefacción** fancoil para calefacción	ventiloconvecteur en mode de chauffage	7.3.2-**4**

	ENGLISH	ESPAÑOL	FRANÇAIS
7.3.2-**5**	air terminal device	**unidad terminal de aire**	bouche d'air
7.3.2-**6**	blending box *mixing box* *mixing unit*	**caja mezcladora** caja de mezcla	boîte de mélange *caisson de mélange* *chambre de mélange* ο
7.3.2-**7**	butterfly damper	**registro de mariposa**	registre papillon
7.3.2-**8**	ceiling diffuser	**difusor de techo**	diffuseur plafonnier
7.3.2-**9**	ceiling outlet	**abertura de techo** boca del techo	bouche de plafond *ouverture de plafond* ο
7.3.2-**10**	coefficient of discharge	**coeficiente de caudal**	coefficient de débit
7.3.2-**11**	core area ο *cross area*	**sección total de la rejilla**	section totale
7.3.2-**12**	cross area *core area* ο	**sección total de la rejilla**	section totale
7.3.2-**13**	damper	**registro** compuerta	registre
7.3.2-**14**	deflector *turning vane*	**deflector** (subst.)	déflecteur
7.3.2-**15**	diffuser	**difusor**	diffuseur
7.3.2-**16**	duct	**conducto** canal	gaine *conduit* ο
7.3.2-**17**	duct distribution	**distribución por conductos**	répartition de l'air par gaines
7.3.2-**18**	duct fittings	**componentes de un conducto**	composants de gaine
7.3.2-**19**	ductwork *trunking (2)* ○	**sistema de conductos**	réseau de gaines *système de gaines* ο
7.3.2-**20**	equal friction method duct sizing	**dimensionado de conductos por el método de igualación de pérdidas de carga**	dimensionnement des conduits par la méthode d'égal frottement
7.3.2-**21**	equalizing damper	**registro de equilibrio**	registre d'égalisation
7.3.2-**22**	exhaust opening	**orificio de salida de aire** boca de salida del aire	bouche de sortie d'air *orifice de sortie d'air* ο
7.3.2-**23**	false ceiling *intermediate ceiling*	**falso techo**	faux plafond
7.3.2-**24**	false floor	**falso suelo** falso piso	faux plancher
7.3.2-**25**	fire-and-smoke damper	**rejilla antifuego**	clapet coupe-feu
7.3.2-**26**	fixed guard	**protección fija**	protecteur fixe
7.3.2-**27**	flexible duct	**conducto flexible**	conduit flexible
7.3.2-**28**	free area	**sección libre de paso del aire**	section libre de passage d'air
7.3.2-**29**	grille *air grille*	**rejilla de aire**	grille à air
7.3.2-**30**	guard	**protección**	protecteur
7.3.2-**31**	in-duct method	**ensayo en conducto**	essai en conduit ο
7.3.2-**32**	induction unit	**eyectoconvector** inductor	éjecto-convecteur ο
7.3.2-**33**	intermediate ceiling *false ceiling*	**falso techo**	faux plafond
7.3.2-**34**	iris damper	**registro de diafragma**	registre à iris

ENGLISH	ESPAÑOL	FRANÇAIS	
linear air diffuser	**difusor de aire lineal**	diffuseur d'air linéaire	7.3.2-**35**
linear grille	**rejilla lineal**	grille linéaire	7.3.2-**36**
louvre	**rejilla**	persienne *louvre (marine)* ◐	7.3.2-**37**
mixing box *blending box* *mixing unit*	**caja mezcladora** caja de mezcla	boîte de mélange *caisson de mélange* *chambre de mélange* ◐	7.3.2-**38**
mixing unit *blending box* *mixing box*	**caja mezcladora** caja de mezcla	boîte de mélange *caisson de mélange* *chambre de mélange* ◐	7.3.2-**39**
modulating damper	**compuerta reguladora**	registre de réglage	7.3.2-**40**
multi-leaf damper	**registro de persianas**	registre à persiennes	7.3.2-**41**
non-ducted air conditioner	**aparato autónomo con descarga directa**	climatiseur non raccordé	7.3.2-**42**
non-return damper	**dispositivo de circulación de aire unidireccional**	registre à sens unique	7.3.2-**43**
nozzle outlet	**tobera de difusión**	buse de diffusion	7.3.2-**44**
outside-air intake duct	**conducto de aire exterior** canal de aire exterior conducto de renovación de aire	gaine de renouvellement d'air *gaine d'air extérieur* ◐	7.3.2-**45**
perforated ceiling *ventilated ceiling* ◐	**techo perforado**	plafond perforé	7.3.2-**46**
plenum chamber *plenum space*	**cámara de distribución de aire** cámara de mezcla de aire	plenum *chambre de répartition d'air* ○	7.3.2-**47**
plenum space *plenum chamber*	**cámara de distribución de aire** cámara de mezcla de aire	plenum *chambre de répartition d'air* ○	7.3.2-**48**
rain louvre *weather louvre* ◐	**rejilla antilluvia**	grille d'air extérieur contre les intempéries	7.3.2-**49**
register ◐	**parrilla de registro**	grille à registre	7.3.2-**50**
shut-off damper	**compuerta de cierre**	registre d'isolement	7.3.2-**51**
single-leaf damper	**registro de persiana**	registre à volet	7.3.2-**52**
slide damper	**registro de guillotina**	registre à glissières *registre (à) guillotine*	7.3.2-**53**
slot diffuser	**difusor lineal**	diffuseur linéaire	7.3.2-**54**
splitter	**álabes directores**	aubage directeur	7.3.2-**55**
static regain method duct sizing	**dimensionado de conductos por el método de reducción de la presión estática**	dimensionnement des conduits par la méthode de regain de pression statique	7.3.2-**56**
trunking (1) ○ *air duct*	**conducto de aire** canal de aire	gaine d'air *conduit d'air* ◐	7.3.2-**57**
trunking (2) ○ *ductwork*	**sistema de conductos**	réseau de gaines *système de gaines* ◐	7.3.2-**58**
turning vane *deflector*	**deflector** (subst.)	déflecteur	7.3.2-**59**
velocity reduction method duct sizing	**dimensionado de conductos por el método de reducción de velocidades**	dimensionnement des conduits par la méthode de réduction des vitesses	7.3.2-**60**
ventilated ceiling ◐ *perforated ceiling*	**techo perforado**	plafond perforé	7.3.2-**61**
weather louvre ◐ *rain louvre*	**rejilla antilluvia**	grille d'air extérieur contre les intempéries	7.3.2-**62**

	ENGLISH	ESPAÑOL	FRANÇAIS
	SECTION 7.3.3 *Air quality:* *specific equipment*	**SUBCAPÍTULO 7.3.3** *Calidad del aire:* *equipos específicos*	**SOUS-CHAPITRE 7.3.3** *Qualité de l'air:* *équipements spécifiques*
7.3.3-**1**	absolute filter	**ultrafiltro** filtro absoluto	filtre absolu *filtre ultrafin* ○ *ultrafiltre* ○
7.3.3-**2**	air cleaner *air filter*	**filtro de aire**	filtre d'air *filtre à air*
7.3.3-**3**	air filter *air cleaner*	**filtro de aire**	filtre d'air *filtre à air*
7.3.3-**4**	automatic roll filter	**filtro de rodillo automático**	filtre à déroulement automatique
7.3.3-**5**	brush filter	**filtro de cepillos**	filtre à brosses
7.3.3-**6**	carbon filter	**filtro de carbón**	filtre à charbon actif
7.3.3-**7**	cartridge filter *cellular filter*	**filtro celular**	filtre à alvéoles *filtre à panneaux* ○ *filtre à cellules* ○
7.3.3-**8**	cellular filter *cartridge filter*	**filtro celular**	filtre à alvéoles *filtre à panneaux* ○ *filtre à cellules* ○
7.3.3-**9**	disposable air filter	**filtro irrecuperable**	filtre jetable *filtre à usage unique* ○
7.3.3-**10**	dry-layer filter	**filtro de capa seca**	filtre sec *filtre à couche sèche* ○
7.3.3-**11**	dust eliminator	**eliminador de polvo**	dépoussiéreur
7.3.3-**12**	dust extracting plant	**instalación de filtrado**	installation de filtrage *installation de dépoussiérage*
7.3.3-**13**	dust-spot procedures for testing air-cleaning devices	**rendimiento opacimétrico**	rendement opacimétrique
7.3.3-**14**	electric precipitator *electrostatic filter* *electrostatic precipitator*	**filtro electrostático**	électrofiltre *filtre électrostatique* *séparateur électrostatique* ○
7.3.3-**15**	electrostatic filter *electric precipitator* *electrostatic precipitator*	**filtro electrostático**	électrofiltre *filtre électrostatique* *séparateur électrostatique* ○
7.3.3-**16**	electrostatic precipitator *electric precipitator* *electrostatic filter*	**filtro electrostático**	électrofiltre *filtre électrostatique* *séparateur électrostatique* ○
7.3.3-**17**	fabric filter	**filtro de tela**	filtre textile
7.3.3-**18**	fibre-pad filter ○ *fibrous filter*	**filtro de fibra** filtro fibroso	filtre à matière fibreuse *filtre à masse fibreuse* ○
7.3.3-**19**	fibrous filter *fibre-pad filter* ○	**filtro de fibra** filtro fibroso	filtre à matière fibreuse *filtre à masse fibreuse* ○
7.3.3-**20**	filter cartridge ○ *filter unit* *filter cell* ○ *filter element* ○	**cartucho filtrante** elemento de un filtro	cartouche filtrante *élément d'un filtre* ○
7.3.3-**21**	filter cell ○ *filter unit* *filter cartridge* ○ *filter element* ○	**cartucho filtrante** elemento de un filtro	cartouche filtrante *élément d'un filtre* ○

ENGLISH	ESPAÑOL	FRANÇAIS	
filter element ○ *filter unit* *filter cartridge* ○ *filter cell* ○	**cartucho filtrante** elemento de un filtro	cartouche filtrante *élément d'un filtre* ○	7.3.3-**22**
filter unit *filter cartridge* ○ *filter cell* ○ *filter element* ○	**cartucho filtrante** elemento de un filtro	cartouche filtrante *élément d'un filtre* ○	7.3.3-**23**
fine filter	**filtro fino**	filtre fin	7.3.3-**24**
fume cupboard	**campana de humos**	hotte (de laboratoire) *sorbonne*	7.3.3-**25**
gravimetric yield	**rendimiento gravimétrico**	rendement gravimétrique	7.3.3-**26**
HEPA filter	**filtro HEPA**	filtre HEPA	7.3.3-**27**
impact filter	**filtro de impacto**	filtre à chocs *filtre à inertie* *séparateur à chocs* ○	7.3.3-**28**
ionizator	**ionizador**	ionisateur	7.3.3-**29**
laminar flow	**flujo laminar**	flux laminaire	7.3.3-**30**
medium-efficacy air filter	**filtro grueso**	filtre grossier	7.3.3-**31**
moving curtain filter ○ *roll filter*	**filtro de rodillo** filtro de cinta sin fin	filtre à déroulement	7.3.3-**32**
ozone	**ozono**	ozone	7.3.3-**33**
ozoniser	**ozonizador**	ozoniseur	7.3.3-**34**
particle meter	**contador de partículas**	compteur de particules	7.3.3-**35**
primary filter	**filtro primario**	préfiltre	7.3.3-**36**
roll filter *moving curtain filter* ○	**filtro de rodillo** filtro de cinta sin fin	filtre à déroulement	7.3.3-**37**
sorption filter	**filtro de sorción**	filtre à sorption	7.3.3-**38**
terminal filter	**filtro final** filtro terminal	filtre terminal	7.3.3-**39**
ULPA filter	**filtro ULPA**	filtre ULPA	7.3.3-**40**
viscous filter	**filtro viscoso**	filtre à imprégnation visqueuse	7.3.3-**41**

SECTION 7.4 *Ventilation*	SUBCAPÍTULO 7.4 *Ventilación*	SOUS-CHAPITRE 7.4 *Ventilation*	
aerofoil (blade) fan	**ventilador de palas aerodinámicas** ventilador de alabes aerodinámicas	ventilateur à aubes profilées	7.4-**1**
axial (flow) fan *propeller fan* ○	**ventilador helicoidal** ventilador axial ventilador axial de baja presión	ventilateur hélicoïde *ventilateur axial* ○	7.4-**2**
backward curved impeller	**rodete con álabes curvados hacia atrás**	roue à aubes tournées vers l'arrière	7.4-**3**
bifurcated fan	**grupo ventilador de "bulbo"** ventilador bifurcado	groupe ventilateur "bulbe" *moto-ventilateur "bulbe"*	7.4-**4**
blade	**paleta** álabe	aube *pale*	7.4-**5**
blower	**ventilador de alta presión** soplante ventilador de impulsión	soufflante	7.4-**6**

	ENGLISH	ESPAÑOL	FRANÇAIS
7.4-**7**	casing	**carcasa** envolvente	enveloppe
7.4-**8**	centrifugal fan	**ventilador centrífugo**	ventilateur centrifuge
7.4-**9**	circulating fan	**ventilador de recirculación**	ventilateur brasseur d'air
7.4-**10**	contra-rotating fan	**ventilador bi-rotatorio**	ventilateur contrarotatif
7.4-**11**	controlled forced-draught ventilation	**ventilación mecánica controlada** ventilación forzada controlada	ventilation mécanique contrôlée
7.4-**12**	cross-flow fan	**ventilador tangencial**	ventilateur tangentiel
7.4-**13**	double-flux controlled forced-draught ventilation	**ventilación mecánica controlada de doble flujo** ventilación forzada controlada de doble flujo	ventilation mécanique contrôlée double flux
7.4-**14**	double inlet fan	**ventilador de doble admisión**	ventilateur à deux ouïes *ventilateur double ouïe* o
7.4-**15**	downstream fairing	**carenado aguas abajo**	carénage aval
7.4-**16**	downstream guide vanes	**distribuidor de salida**	aubage redresseur
7.4-**17**	draught plant	**instalación de ventilación**	système de ventilation
7.4-**18**	ducted fan	**ventilador con conducto(s)**	ventilateur à enveloppe
7.4-**19**	exhauster o *induced draught fan*	**aspirador** extractor ventilador de aspiración	ventilateur d'extraction *ventilateur aspirant* o
7.4-**20**	fan	**ventilador**	ventilateur
7.4-**21**	fan curve	**curva característica de un ventilador**	courbe caractéristique d'un ventilateur
7.4-**22**	fan inlet	**oído**	ouïe d'aspiration du ventilateur
7.4-**23**	fan outlet	**boca de descarga**	ouïe de refoulement du ventilateur
7.4-**24**	fan power	**potencia del ventilador**	puissance du ventilateur
7.4-**25**	forward curved impeller	**rodete con álabes curvados hacia delante**	roue à aubes inclinées vers l'avant
7.4-**26**	gas-tight fan	**ventilador en montaje hermético**	ventilateur étanche
7.4-**27**	guide vane	**álabe director**	aube directrice
7.4-**28**	(guide) vane axial fan	**ventilador de álabes directores**	ventilateur axial à aubage directeur
7.4-**29**	hot-gas fan	**ventilador alta temperatura**	ventilateur pour gaz chauds
7.4-**30**	ignition-protected fan *spark-resistant fan* o	**ventilador antideflagrante**	ventilateur antiétincelles
7.4-**31**	impeller	**rodete**	roue
7.4-**32**	impeller backplate *impeller hub disc* *impeller hub plate*	**disco posterior (del rodete)**	disque arrière (de roue)
7.4-**33**	impeller hub disc *impeller backplate* *impeller hub plate*	**disco posterior (del rodete)**	disque arrière (de roue)
7.4-**34**	impeller hub plate *impeller backplate* *impeller hub disc*	**disco posterior (del rodete)**	disque arrière (de roue)
7.4-**35**	impeller rim o *inlet ring* *impeller shroud* o *wheel cone* o	**disco anterior (del rodete)**	disque avant (de roue) *collerette (de roue)* o

ENGLISH	ESPAÑOL	FRANÇAIS	
impeller shroud ○ *inlet ring* *impeller rim* ○ *wheel cone* ○	**disco anterior (del rodete)**	disque avant (de roue) *collerette (de roue)* ○	7.4-**36**
impeller tip diameter	**diámetro de rodete**	diamètre de la roue	7.4-**37**
induced draught fan *exhauster* ○	**aspirador** extractor ventilador de aspiración	ventilateur d'extraction *ventilateur aspirant* ○	7.4-**38**
industrial fan	**ventilador para usos industriales**	ventilateur industriel	7.4-**39**
inlet box	**caja de aspiración**	caisson d'aspiration *coude d'aspiration* ○	7.4-**40**
inlet ring *impeller rim* ○ *impeller shroud* ○ *wheel cone* ○	**disco anterior (del rodete)**	disque avant (de roue) *collerette (de roue)* ○	7.4-**41**
jet fan	**ventilador en chorro**	ventilateur accélérateur *ventilateur relais* ○	7.4-**42**
mixed-flow fan	**ventilador centrífugo-helicoidal**	ventilateur hélico-centrifuge	7.4-**43**
multistage fan	**ventilador multi-etapa**	ventilateur multiétages	7.4-**44**
paddle-bladed impeller *radial-bladed impeller*	**rodete con álabes radiales**	roue à aubes radiales	7.4-**45**
partition fan	**ventilador de pared**	ventilateur de paroi	7.4-**46**
plate-mounted axial-flow fan	**ventilador axial montado en pletina**	ventilateur hélicoïde monté sur plaque	7.4-**47**
powered roof ventilator	**ventilador de tejado**	tourelle d'extraction *tourelle de ventilation* *ventilateur de toiture* ○	7.4-**48**
propeller fan ○ *axial (flow) fan*	**ventilador helicoidal** ventilador axial ventilador axial de baja presión	ventilateur hélicoïde *ventilateur axial* ○	7.4-**49**
radial-bladed impeller *paddle-bladed impeller*	**rodete con álabes radiales**	roue à aubes radiales	7.4-**50**
reversible axial-flow fan	**ventilador reversible**	ventilateur hélicoïde réversible	7.4-**51**
ring-shaped fan	**ventilador en anillo**	ventilateur annulaire	7.4-**52**
single-flux controlled forced-draught ventilation	**ventilación mecánica controlada de simple flujo** ventilación forzada controlada de simple flujo	ventilation mécanique contrôlée simple flux	7.4-**53**
smoke-ventilating fan	**extractor de humos**	ventilateur de désenfumage	7.4-**54**
spark-resistant fan ○ *ignition-protected fan*	**ventilador antideflagrante**	ventilateur antiétincelles	7.4-**55**
tip clearance	**holgura radial** juego radial	jeu radial	7.4-**56**
tube axial fan ○	**ventilador axial de envolvente**	ventilateur (axial) à enveloppe	7.4-**57**
tubular centrifugal fan	**ventilador en línea**	ventilateur centrifugo-axial	7.4-**58**
upstream fairing	**carenado aguas arriba**	carénage amont	7.4-**59**
upstream guide vanes	**distribuidor de entrada**	distributeur	7.4-**60**
vane axial fan	**ventilador axial con guías**	ventilateur hélicoïde à aubes directrices	7.4-**61**
velocity triangle	**triángulo de las velocidades**	triangle des vitesses	7.4-**62**
ventilator	**aireador**	aérateur	7.4-**63**

	ENGLISH	ESPAÑOL	FRANÇAIS
7.4-**64**	wet-gas fan	**ventilador para gases húmedos**	ventilateur pour gas humides
7.4-**65**	wheel cone ◐ *inlet ring* *impeller rim ◐* *impeller shroud ◐*	**disco anterior (del rodete)**	disque avant (de roue) *collerette (de roue) ◐*

	SECTION 7.5 *Packaged and split* *air-conditioning units*	**SUBCAPÍTULO 7.5** *Instalaciones de acondicionamiento* *de aire monobloc y de elementos* *separados*	**SOUS-CHAPITRE 7.5** *Installations de conditionnement d'air* *monoblocs ou à éléments séparés*
7.5-**1**	air conditioner ◐ *air-conditioning unit*	**acondicionador de aire** aparato de acondicionamiento de aire climatizador	appareil de conditionnement d'air *climatiseur* *conditionneur d'air ◐*
7.5-**2**	air-conditioning unit *air conditioner ◐*	**acondicionador de aire** aparato de acondicionamiento de aire climatizador	appareil de conditionnement d'air *climatiseur* *conditionneur d'air ◐*
7.5-**3**	air-cooled air conditioner	**acondicionador con condensador de aire**	conditionneur d'air à condenseur à air
7.5-**4**	air-handling unit	**unidad de tratamiento de aire**	caisson de traitement d'air
7.5-**5**	air-terminal unit	**unidad terminal**	dispositif terminal
7.5-**6**	all-air system	**sistema "todo aire"**	système "tout air"
7.5-**7**	cassette unit	**cassette**	unité de type cassette
7.5-**8**	central air-conditioning plant	**central de acondicionamiento de aire**	centrale de conditionnement d'air
7.5-**9**	central fan air-conditioning system	**sistema centralizado de acondiciona-miento de aire**	système centralisé de conditionnement d'air
7.5-**10**	chilled beam *cold beam*	**viga fría**	poutre froide *poutre rafraîchissante*
7.5-**11**	cold beam *chilled beam*	**viga fría**	poutre froide *poutre rafraîchissante*
7.5-**12**	console air conditioner ◐	**acondicionamiento (de aire) mural**	conditionneur d'air mural
7.5-**13**	district cooling	**distribución urbana de frío** enfriamiento urbano	refroidissement urbain *distribution urbaine de froid ◐*
7.5-**14**	district heating	**calefacción urbana**	chauffage urbain
7.5-**15**	dual-duct air-conditioning system	**sistema de acondicionamiento de aire de doble conducto**	système de conditionnement d'air à double conduit
7.5-**16**	fan-coil unit *fan-convector unit*	**ventiloconvector**	ventiloconvecteur *batterie ventilée ○*
7.5-**17**	fan-convector unit *fan-coil unit*	**ventiloconvector**	ventiloconvecteur *batterie ventilée ○*
7.5-**18**	four-pipe air-conditioning system	**sistema de acondicionamiento de aire de cuatro tubos**	système de conditionnement d'air à quatre tuyaux
7.5-**19**	free-blow air conditioner *free delivery-type (air-conditioning) unit*	**acondicionador (de aire) de descarga directa** acondicionador (de aire) de impulsión directa	conditionneur d'air à soufflage direct
7.5-**20**	free cooling	**enfriamiento natural**	refroidissement naturel
7.5-**21**	free delivery-type (air-conditioning) unit *free-blow air conditioner*	**acondicionador (de aire) de descarga directa** acondicionador (de aire) de impulsión directa	conditionneur d'air à soufflage direct

ENGLISH	ESPAÑOL	FRANÇAIS	
heat-of-light system ○	**sistema de alumbrado calefactor**	système à éclairage chauffant *système à éclairage intégré*	7.5-**22**
high-pressure air-conditioning plant	**instalación de acondicionamiento de aire a alta presión** instalación de acondicionamiento de aire a gran velocidad	installation de conditionnement d'air à haute pression *installation de conditionnement d'air à grande vitesse*	7.5-**23**
indoor unit	**unidad interior**	unité intérieure	7.5-**24**
low-pressure air-conditioning plant	**instalación de acondicionamiento de aire a baja presión**	installation de conditionnement d'air à basse pression	7.5-**25**
modular (air-conditioning) system	**central modular de acondicionamiento de aire**	installation modulaire (de conditionnement d'air)	7.5-**26**
multisplit air-conditioning system	**multisplit**	multisplit	7.5-**27**
night ventilation	**ventilación nocturna**	ventilation nocturne	7.5-**28**
outdoor unit	**unidad exterior**	unité extérieure	7.5-**29**
packaged air conditioner *self-contained air-conditioning unit* ◑	**climatizador autónomo** acondicionador autónomo (de aire)	conditionneur d'air de type armoire *conditionneur d'air monobloc* ◑	7.5-**30**
regenerative cooling	**enfriamiento por recuperación**	refroidissement par récupération	7.5-**31**
regenerative heating	**calefacción por recuperación**	chauffage par récupération	7.5-**32**
rock-bed regenerative cooling	**enfriamiento por recuperación en lecho de piedras**	refroidissement à récupération sur couches de pierres	7.5-**33**
roof-top (air-conditioning) unit *roof-top conditioner*	**acondicionador (de aire) de techo**	conditionneur d'air en toiture	7.5-**34**
roof-top conditioner *roof-top (air-conditioning) unit*	**acondicionador (de aire) de techo**	conditionneur d'air en toiture	7.5-**35**
room air conditioner (RAC)	**acondicionador de habitaciones**	conditionneur d'air de pièce *conditionneur d'air unitaire*	7.5-**36**
self-contained air-conditioning unit ◑ *packaged air conditioner*	**climatizador autónomo** acondicionador autónomo (de aire)	conditionneur d'air de type armoire *conditionneur d'air monobloc* ◑	7.5-**37**
single-duct air-conditioning system	**sistema de acondicionamiento de aire de un solo conducto**	système de conditionnement d'air à un conduit	7.5-**38**
split (air-conditioning) system	**acondicionador (de aire) con condensador separado** sistema "split"	système split *système (de conditionnement d'air) bibloc* ◑ *conditionneur d'air à condenseur séparé* ○ *conditionneur d'air à deux blocs* ○	7.5-**39**
split unit	**unidad "split"**	split	7.5-**40**
three-pipe air-conditioning system	**sistema de acondicionamiento de aire de tres tubos**	système de conditionnement d'air à trois tuyaux	7.5-**41**
through-the-wall conditioner ◑	**acondicionador (de aire) a través de la pared**	conditionneur d'air "à travers le mur"	7.5-**42**
total energy concept ◑ *total energy system*	**sistema de energía total**	système à énergie totale	7.5-**43**
total energy system *total energy concept* ◑	**sistema de energía total**	système à énergie totale	7.5-**44**
VAV (Variable-Air-Volume) system	**sistema de caudal variable** sistema de volumen variable	système à débit d'air variable (VAV)	7.5-**45**
VRV (Variable-Refrigerant-Volume) system	**sistema de caudal de fluido frigorígeno variable**	système à débit de frigorigène variable) (VRV)	7.5-**46**
water-cooled air conditioner	**acondicionador (de aire) con condensador de agua**	conditionneur d'air à condenseur à eau	7.5-**47**

ENGLISH	ESPAÑOL	FRANÇAIS
7.5-**48** window-air conditioner	**acondicionador (de aire) de ventana**	conditionneur d'air "type fenêtre"
7.5-**49** zone air conditioner	**acondicionador (de aire) de zona**	climatiseur de zone

SECTION 7.6 *Air-conditioned spaces*	**SUBCAPÍTULO 7.6** *Espacios acondicionados*	**SOUS-CHAPITRE 7.6** *Espaces conditionnés*
7.6-**1** climatic chamber *environmental chamber*	**cámara climática**	chambre climatique
7.6-**2** double glazing *dual glazing* ◐	**doble acristalamiento**	double-vitrage
7.6-**3** double window	**doble ventana**	double-fenêtre
7.6-**4** dual glazing ◐ *double glazing*	**doble acristalamiento**	double-vitrage
7.6-**5** enclosed space *enclosure* ◐	**recinto cerrado** espacio cerrado	enceinte
7.6-**6** enclosure ◐ *enclosed space*	**recinto cerrado** espacio cerrado	enceinte
7.6-**7** environmental chamber *climatic chamber*	**cámara climática**	chambre climatique
7.6-**8** fenestration	**superficie acristalada**	surface vitrée
7.6-**9** multiple glazing	**acristalado múltiple**	vitrage multiple
7.6-**10** phytotron	**fitotrón**	phytotron
7.6-**11** shading coefficient	**coeficiente de sombra**	facteur d'écran *facteur d'ombrage*
7.6-**12** shading device	**parasol**	pare-soleil *brise-soleil* ◐
7.6-**13** spot cooling	**enfriamiento localizado**	refroidissement localisé

SECTION 7.7 *Clean rooms*	**SUBCAPÍTULO 7.7** *Cámaras limpias*	**SOUS-CHAPITRE 7.7** *Salles blanches*
7.7-**1** alert level	**nivel de alerta**	niveau d'alerte
7.7-**2** biocontamination	**biocontaminación** contaminación biológica	biocontamination
7.7-**3** classification	**clasificación**	classification
7.7-**4** clean room	**cámara limpia** sala blanca sala limpia	salle blanche *enceinte à empoussiérage contrôlé* *salle propre*
7.7-**5** clean space *clean zone*	**espacio limpio** zona limpia	espace propre *espace à empoussiérage contrôlé*
7.7-**6** clean work station	**puesto de trabajo limpio** estación limpia	poste de travail propre
7.7-**7** clean zone *clean space*	**espacio limpio** zona limpia	espace propre *espace à empoussiérage contrôlé*
7.7-**8** controlled environment	**espacio controlado**	environnement maîtrisé

ENGLISH	ESPAÑOL	FRANÇAIS	
dust-controlled clean rooms: class	**clase de las salas limpias**	classe des locaux à empoussièrement contrôlé	7.7-**9**
fibre	**fibra**	fibre	7.7-**10**
internal generation of particles	**generación interna de partículas**	génération interne de particules	7.7-**11**
macroparticle	**macropartícula**	macroparticule	7.7-**12**
particle	**partícula**	particule	7.7-**13**
particle concentration	**concentración de partículas**	concentration de particules	7.7-**14**
particle size	**tamaño de partícula**	taille de particule	7.7-**15**
particle size distribution	**distribución por tamaño**	distribution granulométrique	7.7-**16**
ultrafine particle	**partícula ultrafina**	particule ultrafine	7.7-**17**
viable particle	**partícula viable**	particule viable	7.7-**18**
zone at risk	**zona de riesgo**	zone à risque	7.7-**19**

Capítulo 8.

BOMBAS DE CALOR

- ● término aceptado
- ○ término obsoleto

ENGLISH	ESPAÑOL	FRANÇAIS	
CHAPTER 8 *Heat pumps*	**CAPÍTULO 8** *Bombas de calor*	**CHAPITRE 8** *Pompes à chaleur*	
absorption heat pump	**bomba de calor de absorción**	pompe à chaleur à absorption	8-1
adsorption heat pump	**bomba de calor de adsorción**	pompe à chaleur à adsorption	8-2
air-source heat pump	**bomba de calor "aire-xxx"** bomba de calor que usa aire exterior como foco caliente	pompe à chaleur sur l'air	8-3
air-to-air heat pump	**bomba de calor aire-aire**	pompe à chaleur air-air	8-4
air-to-water heat pump	**bomba de calor aire-agua**	pompe à chaleur air-eau	8-5
brine-to-air heat pump ⊙ *water-to-air heat pump*	**bomba de calor agua-aire** bomba de calor salmuera-aire	pompe à chaleur eau-air	8-6
brine-to-water heat pump	**bomba de calor salmuera-agua**	pompe à chaleur saumure-eau	8-7
chemical heat pump	**bomba de calor química**	pompe à chaleur à réaction chimique	8-8
closed-loop ground-source heat pump ⊙ *ground-source heat pump* *ground-coupled heat pump* *geothermal heat pump* ⊙ *ground-loop heat pump* ⊙	**bomba de calor acoplada a la tierra** bomba de calor que usa la tierra como foco frío o caliente	pompe à chaleur couplée au sol	8-9
coefficient of performance (of a heat pump)	**coeficiente de prestación**	coefficient de performance (d'une pompe à chaleur) *coefficient d'efficacité calorifique (d'une pompe à chaleur)* ⊙ *coefficient d'amplification (d'une pompe à chaleur)* ⊙	8-10
compression heat pump	**bomba de calor de compresión**	pompe à chaleur à compression	8-11
direct-expansion ground-coupled heat pump	**bomba de calor de expansión directa sobre tierra**	pompe à chaleur sol-air ou sol-eau à évaporation directe	8-12
geothermal heat pump ⊙ *ground-source heat pump* *ground-coupled heat pump* *closed-loop ground-source heat pump* ⊙ *ground-loop heat pump* ⊙	**bomba de calor acoplada a la tierra** bomba de calor que usa la tierra como foco frío o caliente	pompe à chaleur couplée au sol	8-13
ground-coupled heat pump *ground-source heat pump* *closed loop ground-source heat pump* ⊙ *geothermal heat pump* ⊙ *ground-loop heat pump* ⊙	**bomba de calor acoplada a la tierra** bomba de calor que usa la tierra como foco frío o caliente	pompe à chaleur couplée au sol	8-14
ground-loop heat pump ⊙ *ground-source heat pump* *ground-coupled heat pump* *closed-loop ground-source heat pump* ⊙ *geothermal heat pump* ⊙	**bomba de calor acoplada a la tierra** bomba de calor que usa la tierra como foco frío o caliente	pompe à chaleur couplée au sol	8-15
ground-source heat pump *ground-coupled heat pump* *closed-loop ground-source heat pump* ⊙ *geothermal heat pump* ⊙ *ground-loop heat pump* ⊙	**bomba de calor acoplada a la tierra** bomba de calor que usa la tierra como foco frío o caliente	pompe à chaleur couplée au sol	8-16
ground-water heat pump ⊙	**bomba de calor "tierra-agua"**	pompe à chaleur sur eau souterraine	8-17
heat output	**calor entregado (por el sistema)**	puissance thermique	8-18
heat pump	**bomba de calor**	pompe à chaleur	8-19

	ENGLISH	ESPAÑOL	FRANÇAIS
8-20	heat pump boiler ○ *heat pump water heater*	**bomba de calor para producción de agua caliente** bomba de calor para producción de agua caliente sanitaria	pompe à chaleur pour chauffage d'eau
8-21	heat pump water heater *heat pump boiler ○*	**bomba de calor para producción de agua caliente** bomba de calor para producción de agua caliente sanitaria	pompe à chaleur pour chauffage d'eau
8-22	heat-recovery heat pump	**bomba de calor de recuperación de calor**	pompe à chaleur pour récupération de chaleur
8-23	heat transformer *temperature amplifier ○*	**amplificador de temperatura**	transformateur de chaleur
8-24	heating energy	**energía calorífica**	énergie thermique
8-25	heating seasonal performance factor *HSPF*	**coeficiente de prestación estacional**	coefficient de performance moyen saisonnier
8-26	HSPF *heating seasonal performance factor*	**coeficiente de prestación estacional**	coefficient de performance moyen saisonnier
8-27	mechanical vapour recompression	**bomba de calor con motor a vapor**	recompression mécanique de vapeur
8-28	primary energy ratio	**coeficiente de prestación primario**	efficacité rapportée à l'énergie primaire
8-29	reverse-cycle heating ○ *thermodynamic heating ○*	**calefacción termodinámica**	chauffage par cycle inversé ○ *chauffage par cycle frigorifique ○* *chauffage thermodynamique ○*
8-30	solar-assisted heat pump ○	**bomba de calor con apoyo solar**	pompe à chaleur assistée par l'énergie solaire
8-31	surface-water heat pump ○	**bomba de calor por agua superficial**	pompe à chaleur sur eau de surface
8-32	temperature amplifier ○ *heat transformer*	**amplificador de temperatura**	transformateur de chaleur
8-33	thermal vapour recompression	**bomba de calor con motor a vapor**	thermocompression de vapeur
8-34	thermodynamic heating ○ *reverse cycle heating ○*	**calefacción termodinámica**	chauffage par cycle inversé ○ *chauffage par cycle frigorifique ○* *chauffage thermodynamique ○*
8-35	thermoelectric heat pump	**bomba de calor termoeléctrica**	pompe à chaleur thermoélectrique
8-36	ventilation air heat pump ○	**bomba de calor por aire de ventilación**	pompe à chaleur sur air extrait
8-37	water-loop heat pump	**bomba de calor en ciclo de agua**	pompe à chaleur sur boucle d'eau
8-38	water-source heat pump	**bomba de calor "agua-xxx"** bomba de calor que usa agua como foco caliente	pompe à chaleur sur eau
8-39	water-to-air heat pump *brine-to-air heat pump ○*	**bomba de calor agua-aire** bomba de calor salmuera-aire	pompe à chaleur eau-air
8-40	water-to-water heat pump	**bomba de calor agua-agua**	pompe à chaleur eau-eau

Capítulo 9. | Criología

◑ término aceptado

○ término obsoleto

ENGLISH	ESPAÑOL	FRANÇAIS	
SECTION 9.1 *Cryophysics* **SECTION 9.1.1** *Cryogenics and cryoengineering*	**SUBCAPÍTULO 9.1** *Criofísica* **SUBCAPÍTULO 9.1.1** *Criogenia y criotecnía*	**SOUS-CHAPITRE 9.1** *Cryophysique* **SOUS-CHAPITRE 9.1.1** *Cryogénie et cryotechnique*	
adiabatic demagnetization	**desimantación adiabática**	désaimantation adiabatique	9.1.1-**1**
air fractionation ○ *air separation*	**separación del aire**	séparation de l'air	9.1.1-**2**
air separation *air fractionation* ○	**separación del aire**	séparation de l'air	9.1.1-**3**
bubble chamber	**cámara de burbujas**	chambre à bulles	9.1.1-**4**
cold box	**caja fría**	boîte froide	9.1.1-**5**
cryoalternator	**crioalternador**	cryoalternateur	9.1.1-**6**
cryocable	**criocable**	cryocâble	9.1.1-**7**
cryochemistry	**crioquímica**	cryochimie	9.1.1-**8**
cryoconductor	**crioconductor** (subst.)	cryoconducteur (subst.) *hyperconducteur*	9.1.1-**9**
cryocooling *cryogenic cooling*	**enfriamiento criogénico** crioenfriamiento	cryorefroidissement	9.1.1-**10**
cryoelectric *cryoelectrical*	**crioeléctrico** (adj.)	cryoélectrique	9.1.1-**11**
cryoelectrical *cryoelectric*	**crioeléctrico** (adj.)	cryoélectrique	9.1.1-**12**
cryoelectronics	**crioelectrónica**	cryoélectronique	9.1.1-**13**
cryoelectrotechnics	**crioelectrotecnia**	cryoélectrotechnique	9.1.1-**14**
cryoengineering *cryogenic engineering* ◑	**criotecnia**	cryotechnique *ingénierie cryogénique* *technique cryogénique*	9.1.1-**15**
cryogen *cryogenic fluid*	**fluido criogénico** criógeno	cryogène *fluide cryogénique*	9.1.1-**16**
cryogenic	**criogénico** (adj.)	cryogénique	9.1.1-**17**
cryogenic bath	**baño criogénico**	bain cryogénique	9.1.1-**18**
cryogenic cooling *cryocooling*	**enfriamiento criogénico** crioenfriamiento	cryorefroidissement	9.1.1-**19**
cryogenic engineering ◑ *cryoengineering*	**criotecnia**	cryotechnique *ingénierie cryogénique* *technique cryogénique*	9.1.1-**20**
cryogenic equipment	**equipo criogénico** aparato criogénico material criogénico	matériel cryogénique *appareil cryogénique* *équipement cryogénique* ○	9.1.1-**21**
cryogenic fluid *cryogen*	**fluido criogénico** criógeno	cryogène *fluide cryogénique*	9.1.1-**22**
cryogenic liquid	**líquido criogénico**	liquide cryogénique	9.1.1-**23**
cryogenic plant	**instalación criogénica**	installation cryogénique	9.1.1-**24**
cryogenic process	**procedimiento criogénico**	procédé cryogénique	9.1.1-**25**
cryogenic propellant ◑ *cryopropellant*	**criopropulsor**	ergol cryogénique	9.1.1-**26**

ENGLISH	ESPAÑOL	FRANÇAIS
9.1.1-**27**	cryogenic refrigerator ◐ *cryorefrigerator*	**criorrefrigerador** refrigerador criogénico	cryoréfrigérateur
9.1.1-**28**	cryogenic storage	**almacenamiento criogénico**	stockage cryogénique
9.1.1-**29**	cryogenic (storage) vessel	**depósito criogénico**	réservoir cryogénique
9.1.1-**30**	cryogenic tanker	**buque criogénico**	citerne cryogénique
9.1.1-**31**	cryogenic technique ◐ *cryotechnique*	**criotécnica**	cryotechnique (expérimentale)
9.1.1-**32**	cryogenic technology ◐ *cryotechnology*	**criotecnología**	cryotechnique (procédés)
9.1.1-**33**	cryogenic valve ◐ *cryovalve*	**crioválvula**	cryovanne *vanne cryogénique*
9.1.1-**34**	cryogenics *cryology*	**criología**	cryogénie
9.1.1-**35**	cryoliquefier	**criolicuefactor** criolicuador	cryoliquéfacteur
9.1.1-**36**	cryology *cryogenics*	**criología**	cryogénie
9.1.1-**37**	cryomachining	**criomecanizado** (subst.)	cryo-usinage
9.1.1-**38**	cryomagnet	**crioimán**	cryoaimant
9.1.1-**39**	cryomicroscope	**criomicroscopio**	cryomicroscope
9.1.1-**40**	cryomotor	**criomotor**	cryomoteur
9.1.1-**41**	cryophysics	**criofísica**	cryophysique
9.1.1-**42**	cryopropellant *cryogenic propellant* ◐	**criopropulsor**	ergol cryogénique
9.1.1-**43**	cryopump	**criobomba**	cryopompe
9.1.1-**44**	cryorefrigerator *cryogenic refrigerator* ◐	**criorrefrigerador** refrigerador criogénico	cryoréfrigérateur
9.1.1-**45**	cryoresistive	**crioconductor** (adj.)	cryoconducteur (adj.)
9.1.1-**46**	cryostat	**criostato**	cryostat
9.1.1-**47**	cryotechnique *cryogenic technique* ◐	**criotécnica**	cryotechnique (expérimentale)
9.1.1-**48**	cryotechnology *cryogenic technology* ◐	**criotecnología**	cryotechnique (procédés)
9.1.1-**49**	cryotemperature	**criotemperatura**	cryotempérature *température cryogénique*
9.1.1-**50**	cryotransformer	**criotransformador** (subst.)	cryotransformateur
9.1.1-**51**	cryotrap	**criointerceptor**	cryopiège
9.1.1-**52**	cryotron ○	**criotrón**	cryotron
9.1.1-**53**	cryovalve *cryogenic valve* ◐	**crioválvula**	cryovanne *vanne cryogénique*
9.1.1-**54**	Dewar (vessel) *vacuum flask* ○	**vaso Dewar**	(vase) Dewar
9.1.1-**55**	dilution refrigerator	**refrigerador de dilución**	réfrigérateur à dilution
9.1.1-**56**	expansion method *Simon's expansion method*	**crioenfriamiento por expansión**	cryorefroidissement par détente *détente de Simon*
9.1.1-**57**	gas separation unit	**aparato separador de gases**	appareil de séparation des gaz

ENGLISH	ESPAÑOL	FRANÇAIS	
helium desorption method	**crioenfriamiento por desorción**	cryorefroidissement par désorption	9.1.1-**58**
magnetic cooling ⦿	**enfriamiento magnético**	refroidissement magnétique	9.1.1-**59**
nuclear alignment	**alineación nuclear**	alignement nucléaire	9.1.1-**60**
nuclear cooling	**enfriamiento nuclear**	refroidissement nucléaire	9.1.1-**61**
nuclear orientation	**orientación nuclear**	orientation nucléaire	9.1.1-**62**
nuclear polarization	**polarización nuclear**	polarisation nucléaire	9.1.1-**63**
phonon drag	**arrastre de (los) fonomes**	entraînement des phonons	9.1.1-**64**
Simon's expansion method *expansion method*	**crioenfriamiento por expansión**	cryorefroidissement par détente *détente de Simon*	9.1.1-**65**
supercritical	**supercrítico** (adj.)	supercritique	9.1.1-**66**
vacuum flask ○ *Dewar (vessel)*	**vaso Dewar**	(vase) Dewar	9.1.1-**67**

SECTION 9.1.2 *Liquid helium*	SUBCAPÍTULO 9.1.2 *Helio líquido*	SOUS-CHAPITRE 9.1.2 *Hélium liquide*	
creep rate (He II)	**velocidad de arrastre**	vitesse d'écoulement en film *vitesse de grimpage* ⦿	9.1.2-**1**
critical velocity (He II)	**velocidad crítica (He II)**	vitesse critique (He II)	9.1.2-**2**
fountain effect (He II) *thermomechanical effect* ⦿	**efecto fuente**	effet fontaine *effet thermomécanique* ⦿	9.1.2-**3**
fourth sound	**cuarto sonido**	quatrième son	9.1.2-**4**
helium film (He II) *Rollin film* ⦿	**película de helio**	film d'hélium *film de Rollin* ⦿	9.1.2-**5**
lambda leak (He II) ⦿	**fuga lambda**	fuite lambda	9.1.2-**6**
lambda line (He-4)	**curva lambda**	courbe lambda (He-4)	9.1.2-**7**
lambda point (He-4)	**punto lambda**	point lambda (He-4)	9.1.2-**8**
parafluidity ○ *parasuperfluidity*	**parasuperfluidez**	parasuperfluidité *parafluidité* ○	9.1.2-**9**
parasuperfluidity *parafluidity* ○	**parasuperfluidez**	parasuperfluidité *parafluidité* ○	9.1.2-**10**
Rollin film ⦿ *helium film (He II)*	**película de helio**	film d'hélium *film de Rollin* ⦿	9.1.2-**11**
rotons	**rotones**	rotons	9.1.2-**12**
second sound (He II)	**segundo sonido**	deuxième son	9.1.2-**13**
superfluid flow (He II)	**corriente superfluida** flujo superfluido	écoulement superfluide	9.1.2-**14**
superfluidity (He II)	**superfluidez**	superfluidité	9.1.2-**15**
superleak (He II)	**superfuga**	superfuite	9.1.2-**16**
thermomechanical effect ⦿ *fountain effect (He II)*	**efecto fuente**	effet fontaine *effet thermomécanique* ⦿	9.1.2-**17**
third sound	**tercer sonido**	troisième son	9.1.2-**18**
two-fluid model (He II)	**modelo de dos fluidos**	modèle à deux fluides	9.1.2-**19**
zeroth sound	**sonido cero**	son zéro	9.1.2-**20**

SECTION 9.1.3 *Superconductivity*	**SUBCAPÍTULO 9.1.3** *Superconductividad*	**SOUS-CHAPITRE 9.1.3** *Supraconductivité*
9.1.3-1 AC Josephson effect	**efecto Josephson alterno**	effet Josephson alternatif
9.1.3-2 adiabatic stabilization	**estabilización adiabática**	stabilisation adiabatique
9.1.3-3 coherence length	**longitud de coherencia**	longueur de cohérence
9.1.3-4 critical current	**corriente crítica**	courant critique
9.1.3-5 critical field (H_c) *thermodynamical critical field* ○	**campo crítico (H_c)**	champ critique (H_c) *champ critique thermodynamique* ○
9.1.3-6 critical temperature (superconductor)	**temperatura crítica (superconducción)**	température critique (supraconducteur)
9.1.3-7 cryostabilization	**crioestabilización**	cryostabilisation *stabilisation cryogénique* ○
9.1.3-8 DC Josephson effect	**efecto Josephson continuo**	effet Josephson continu
9.1.3-9 degradation (superconductor)	**degradación**	dégradation (supraconducteur)
9.1.3-10 dynamic stabilization	**estabilización dinámica**	stabilisation dynamique
9.1.3-11 filamentary superconductor	**superconductor de filamentos múltiples**	supraconducteur filamentaire
9.1.3-12 flux flow	**corriente de flujo (magnético)**	écoulement de flux
9.1.3-13 flux jump	**salto de flujo (magnético)**	saut de flux
9.1.3-14 fluxoid	**fluxoide**	fluxoïde
9.1.3-15 fluxon	**fluxón**	fluxon
9.1.3-16 frozen-in flux ○ *trapped flux*	**flujo remanente**	flux piégé
9.1.3-17 intermediate state	**estado intermedio**	état intermédiaire
9.1.3-18 intrinsic stabilization	**estabilización intrínseca**	stabilisation intrinsèque
9.1.3-19 Josephson effects	**efecto Josephson**	effets Josephson
9.1.3-20 lower critical field (H_{c1})	**campo crítico inferior (H_{c1})**	champ critique inférieur (H_{c1})
9.1.3-21 magnetic penetration depth *penetration depth* ○	**profundidad de penetración**	profondeur de pénétration
9.1.3-22 maximum recovery current *minimum propagating current*	**corriente mínima de propagación** corriente máxima de recuperación corriente máxima de regeneración	courant minimal de propagation résistive *courant maximal de récupération*
9.1.3-23 Meissner state	**estado de Meissner**	état de Meissner
9.1.3-24 minimum propagating current *maximum recovery current*	**corriente mínima de propagación** corriente máxima de recuperación corriente máxima de regeneración	courant minimal de propagation résistive *courant maximal de récupération*
9.1.3-25 minimum propagating zone *MPZ*	**zona mínima de propagación**	zone résistive minimale de propagation
9.1.3-26 mixed state	**estado mixto**	état mixte
9.1.3-27 MPZ *minimum propagating zone*	**zona mínima de propagación**	zone résistive minimale de propagation
9.1.3-28 normal state	**estado normal**	état normal
9.1.3-29 penetration depth ○ *magnetic penetration depth*	**profundidad de penetración**	profondeur de pénétration
9.1.3-30 persistent current	**corriente persistente**	courant persistant
9.1.3-31 proximity effect	**efecto de proximidad**	effet de proximité

ENGLISH	ESPAÑOL	FRANÇAIS	
SQUID (superconducting quantum interference device)	**interferómetro cuántico super-conductor** "squid"	SQUID *interféromètre quantique supraconducteur*	9.1.3-**32**
stabilization	**estabilización**	stabilisation	9.1.3-**33**
superconducting state *superconductive state* ○	**estado superconductor**	état supraconducteur	9.1.3-**34**
superconduction ○ *superconductivity*	**superconducción**	supraconductivité *supraconduction*	9.1.3-**35**
superconductive state ○ *superconducting state*	**estado superconductor**	état supraconducteur	9.1.3-**36**
superconductivity *superconduction* ○	**superconducción**	supraconductivité *supraconduction*	9.1.3-**37**
superconductor	**superconductor**	supraconducteur	9.1.3-**38**
thermodynamical critical field ○ *critical field (H$_c$)*	**campo crítico (H$_c$)**	champ critique (H$_c$) *champ critique thermodynamique* ○	9.1.3-**39**
transition temperature	**temperatura de transición**	température de transition	9.1.3-**40**
trapped flux *frozen-in flux* ○	**flujo remanente**	flux piégé	9.1.3-**41**
type I superconductor	**superconductor de tipo I**	supraconducteur de type I	9.1.3-**42**
type II superconductor	**superconductor de tipo II**	supraconducteur de type II	9.1.3-**43**
upper critical field (H$_{c2}$)	**campo crítico superior (H$_{c2}$)**	champ critique supérieur (H$_{c2}$)	9.1.3-**44**

SECTION 9.1.4 *Liquefied gases*	SUBCAPÍTULO 9.1.4 *Gases licuados*	SOUS-CHAPITRE 9.1.4 *Gaz liquéfiés*	
air-separation plant	**instalación de separación de aire**	installation de séparation d'air	9.1.4-**1**
ambient air vaporizer	**vaporizador de ambiente**	vaporiseur dans l'air ambiant *vaporisateur dans l'air ambiant*	9.1.4-**2**
argon column *argon side column* ○ *side-arm column* ○	**columna de argón**	colonne d'argon	9.1.4-**3**
argon side column ○ *argon column* *side-arm column* ○	**columna de argón**	colonne d'argon	9.1.4-**4**
balance stream ○ *trumpler pass*	**caudal de equilibrado** flujo de equilibrado	flux d'équilibrage	9.1.4-**5**
ballasting	**balastado**	ballastage	9.1.4-**6**
base load (LNG plant)	**planta base**	capacité de production nominale (installation de GNL)	9.1.4-**7**
bayonet joint	**unión de bayoneta** racor de bayoneta	raccord à baïonnette	9.1.4-**8**
boil-off	**vaporizado** (subst.)	pertes par évaporation	9.1.4-**9**
boil-off rate (BOR) *BOR* *evaporation rate* *NER* *net evaporation rate (NER)*	**tasa de evaporación**	taux d'évaporation *taux de pertes par évaporation*	9.1.4-**10**

	ENGLISH	ESPAÑOL	FRANÇAIS
9.1.4-11	BOR *boil-off rate (BOR)* *evaporation rate* *NER* *net evaporation rate (NER)*	**tasa de evaporación**	taux d'évaporation *taux de pertes par évaporation*
9.1.4-12	brazed-aluminium heat exchanger *plate-and-fin heat exchanger* ○	**intercambiador de calor de aluminio soldado**	échangeur de chaleur en aluminium brasé
9.1.4-13	bund wall	**muro de contención**	cuvette de rétention *mur de rétention* ○
9.1.4-14	cascade cycle	**ciclo en cascada**	cycle à cascade *procédé à cascade* ○
9.1.4-15	Claude cycle ○ *medium-pressure cycle*	**ciclo de media presión**	cycle moyenne pression
9.1.4-16	CLOX *crude liquid oxygen (CLOX)*	**oxígeno líquido bruto**	oxygène liquide brut
9.1.4-17	cold recovery	**recuperación de frío**	récupération de froid
9.1.4-18	compander	**expansor-compresor**	unité de compression/détente
9.1.4-19	condenser-reboiler	**condensador-rehervidor**	condenseur-vaporiseur
9.1.4-20	crude liquid oxygen (CLOX) *CLOX*	**oxígeno líquido bruto**	oxygène liquide brut
9.1.4-21	cryogenic storage tank *vacuum-insulated tank*	**recipiente con aislamiento al vacío**	réservoir isolé sous vide
9.1.4-22	customer station *vaporization station* *satellite station*	**instalación de regasificación** instalación de vaporización	poste de vaporisation
9.1.4-23	denitrogenation	**desnitrogenación**	dénitrogénation *désazotation*
9.1.4-24	double-column plant	**planta de doble columna**	unité à double colonne
9.1.4-25	downflow condenser-reboiler	**condensador-rehervidor de flujo descendente**	condenseur-vaporiseur à flux descendant
9.1.4-26	EOR (enhanced oil recovery)	**recuperación asistida**	récupération assistée
9.1.4-27	evaporation rate *boil-off rate (BOR)* *BOR* *NER* *net evaporation rate (NER)*	**tasa de evaporación**	taux d'évaporation *taux de pertes par évaporation*
9.1.4-28	expander ○ *expansion turbine*	**turbina de expansión**	turbine à expansion *turbine de détente*
9.1.4-29	expansion bellows	**fuelle compensador (de expansión)**	soufflet d'expansion
9.1.4-30	expansion engine	**motor de expansión**	moteur à expansion
9.1.4-31	expansion turbine *expander* ○	**turbina de expansión**	turbine à expansion *turbine de détente*
9.1.4-32	flat-bottom tank	**recipiente de fondo plano**	réservoir à fond plat
9.1.4-33	GAN *gaseous nitrogen (GAN)*	**nitrógeno gas**	azote gazeux
9.1.4-34	gaseous nitrogen (GAN) *GAN*	**nitrógeno gas**	azote gazeux
9.1.4-35	gaseous oxygen (GOX) *GOX*	**oxígeno gas**	oxygène gazeux
9.1.4-36	GOX *gaseous oxygen (GOX)*	**oxígeno gas**	oxygène gazeux

ENGLISH	ESPAÑOL	FRANÇAIS	
heat leakage	**pérdidas de calor**	fuite thermique	9.1.4-**37**
	fuga térmica	*déperdition de chaleur* ○	
Heylandt cycle	**ciclo de Heylandt**	cycle de Heylandt	9.1.4-**38**
high-pressure cycle			
high-pressure column	**columna de alta presión**	colonne haute pression	9.1.4-**39**
HP column ○			
lower column ○			
high-pressure cycle	**ciclo de Heylandt**	cycle de Heylandt	9.1.4-**40**
Heylandt cycle			
HP column ○	**columna de alta presión**	colonne haute pression	9.1.4-**41**
high-pressure column			
lower column ○			
inner cascade cycle	**ciclo de fluido frigorígeno mixto**	cycle à fluide frigorigène mixte	9.1.4-**42**
mixed-refrigerant cycle			
mixed-refrigerant liquefier			
internal-compression plant ○	**planta de bombeo de líquido**	unité de séparation d'air à pompe	9.1.4-**43**
liquid-pump plant		*unité de séparation d'air à compression interne* ○	
Kellog cycle ○	**ciclo de baja presión**	cycle basse pression	9.1.4-**44**
low-pressure cycle			
low-pressure liquefier			
Le Rouget cycle	**ciclo de Le Rouget**	cycle de Le Rouget	9.1.4-**45**
LIN	**nitrógeno líquido**	azote liquide	9.1.4-**46**
liquid nitrogen (LIN)			
Linde column ○	**planta de columna unica**	unité à une seule colonne	9.1.4-**47**
single-column plant			
Linde cycle	**ciclo de Linde**	cycle de Linde	9.1.4-**48**
Linde-Frankl process	**proceso Linde-Frank**	procédé Linde-Frankl	9.1.4-**49**
Linde high-pressure cycle	**ciclo de Linde alta presión**	cycle de Linde haute pression	9.1.4-**50**
liquefied natural gas (LNG)	**gas natural licuado**	gaz naturel liquéfié	9.1.4-**51**
LNG	GNL	*GNL*	
liquefied petroleum gas (LPG)	**gas de petróleo licuado**	gaz de pétrole liquéfié (GPL)	9.1.4-**52**
LPG		*GPL*	
liquid nitrogen (LIN)	**nitrógeno líquido**	azote liquide	9.1.4-**53**
LIN			
liquid nitrogen wash	**lavado con nitrógeno líquido**	lavage à l'azote liquide	9.1.4-**54**
liquid oxygen (LOX)	**oxígeno líquido**	oxygène liquide	9.1.4-**55**
LOX			
liquid-pump plant	**planta de bombeo de líquido**	unité de séparation d'air à pompe	9.1.4-**56**
internal-compression plant ○		*unité de séparation d'air à compression interne* ○	
LNG	**gas natural licuado**	gaz naturel liquéfié	9.1.4-**57**
liquefied natural gas (LNG)	GNL	*GNL*	
LNG ship of a membrane type	**metanero del tipo membrana**	bateau GNL du type membrane	9.1.4-**58**
		méthanier du type membrane ○	
LNG ship of a Moss type	**metanero del tipo Moss**	bateau GNL du type Moss	9.1.4-**59**
Moss-Rosenberg-type ship ○	buque GNL del tipo Moss	*méthanier du type Moss* ○	
LNG ship of a prismatic type	**metanero de cubas prismáticas**	bateau GNL du type prismatique	9.1.4-**60**
		méthanier à cuves prismatiques ○	
loading arm	**brazo de carga**	bras de chargement	9.1.4-**61**

	ENGLISH	ESPAÑOL	FRANÇAIS
9.1.4-62	low-pressure air separation plant *low-pressure plant* o	**unidad de separación de aire de baja presión**	unité de séparation d'air basse pression
9.1.4-63	low-pressure column *upper column* o	**columna de baja presión**	colonne basse pression
9.1.4-64	low-pressure cycle *low-pressure liquefier* *Kellog cycle* o	**ciclo de baja presión**	cycle basse pression
9.1.4-65	low-pressure liquefier *low-pressure cycle* *Kellog cycle* o	**ciclo de baja presión**	cycle basse pression
9.1.4-66	low-pressure plant o *low-pressure air-separation plant*	**unidad de separación de aire de baja presión**	unité de séparation d'air basse pression
9.1.4-67	lower column o *high-pressure column* *HP column* o	**columna de alta presión**	colonne haute pression
9.1.4-68	LOX *liquid oxygen (LOX)*	**oxígeno líquido**	oxygène liquide
9.1.4-69	LPG *liquefied petroleum gas (LPG)*	**gas de petróleo licuado**	gaz de pétrole liquéfié (GPL) *GPL*
9.1.4-70	main heat exchanger	**cambiador de calor principal**	échangeur de chaleur principal
9.1.4-71	medium-pressure air-separation plant *medium-pressure plant* o	**unidad de separación de aire de media presión**	unité de séparation d'air moyenne pression
9.1.4-72	medium-pressure cycle *Claude cycle* o	**ciclo de media presión**	cycle moyenne pression
9.1.4-73	medium-pressure plant o *medium-pressure air-separation plant*	**unidad de separación de aire de media presión**	unité de séparation d'air moyenne pression
9.1.4-74	methane wash process	**proceso de lavado al metano**	procédé de lavage au méthane
9.1.4-75	mixed-component refrigerant o *mixed refrigerant* *multi-component refrigerant* o	**fluido frigorígeno mixto** mezcla de refrigerantes	fluide frigorigène mixte
9.1.4-76	mixed refrigerant *mixed-component refrigerant* o *multi-component refrigerant* o	**fluido frigorígeno mixto** mezcla de refrigerantes	fluide frigorigène mixte
9.1.4-77	mixed-refrigerant cycle *inner cascade cycle* *mixed-refrigerant liquefier*	**ciclo de fluido frigorígeno mixto**	cycle à fluide frigorigène mixte
9.1.4-78	mixed-refrigerant liquefier *mixed-refrigerant cycle* *inner cascade cycle*	**ciclo de fluido frigorígeno mixto**	cycle à fluide frigorigène mixte
9.1.4-79	MLI *multilayer insulation (MLI)* *superinsulation*	**aislamiento reforzado**	superisolation
9.1.4-80	molecular sieve	**tamiz molecular**	tamis moléculaire
9.1.4-81	Moss-Rosenberg-type ship o *LNG ship of a Moss type*	**metanero del tipo Moss** buque GNL del tipo Moss	bateau GNL du type Moss *méthanier du type Moss* o
9.1.4-82	multi-component refrigerant o *mixed refrigerant* *mixed-component refrigerant* o	**fluido frigorígeno mixto** mezcla de refrigerantes	fluide frigorigène mixte
9.1.4-83	multilayer insulation (MLI) *superinsulation* *MLI*	**aislamiento reforzado**	superisolation

ENGLISH	ESPAÑOL	FRANÇAIS	
natural gas (NG) *NG*	**gas natural**	gaz naturel	9.1.4-**84**
natural gas liquid (NGL) *NGL*	**gas natural licuado**	gaz naturel liquide	9.1.4-**85**
NER *boil-off rate (BOR)* *BOR* *evaporation rate* *net evaporation rate (NER)*	**tasa de evaporación**	taux d'évaporation *taux de pertes par évaporation*	9.1.4-**86**
net evaporation rate (NER) *boil-off rate (BOR)* *BOR* *evaporation rate* *NER*	**tasa de evaporación**	taux d'évaporation *taux de pertes par évaporation*	9.1.4-**87**
NG *natural gas (NG)*	**gas natural**	gaz naturel	9.1.4-**88**
NGL *natural gas liquid (NGL)*	**gas natural licuado**	gaz naturel liquide	9.1.4-**89**
nitrogen generator	**generador de nitrógeno**	générateur d'azote	9.1.4-**90**
open-rack vaporizer	**regasificador a corriente de agua**	regazéifieur à ruissellement d'eau *vaporiseur à ruissellement d'eau*	9.1.4-**91**
oxygen generator	**generador de oxígeno**	générateur d'oxygène	9.1.4-**92**
para- or orthohydrogen	**para- o ortohidrógeno**	para- ou orthohydrogène	9.1.4-**93**
peak-lopping plant ◑ *peak-shave (LNG) plant* *peak-shaving plant*	**central eliminadora de picos (planta de gas natural licuado)**	station d'écrètement de pointe	9.1.4-**94**
peak-shave (LNG) plant *peak-shaving plant* *peak-lopping plant ◑*	**central eliminadora de picos (planta de gas natural licuado)**	station d'écrètement de pointe	9.1.4-**95**
peak-shaving plant *peak-shave (LNG) plant* *peak-lopping plant ◑*	**central eliminadora de picos (planta de gas natural licuado)**	station d'écrètement de pointe	9.1.4-**96**
plate-and-fin heat exchanger ◑ *brazed-aluminium exchanger*	**intercambiador de calor de aluminio soldado**	échangeur de chaleur en aluminium brasé	9.1.4-**97**
poor liquid *shelf liquid ○*	**líquido pobre**	liquide pauvre	9.1.4-**98**
rare gases	**gases raros** gases inertes	gaz rares	9.1.4-**99**
recondenser	**recondensador** retrocondensador	recondenseur *reliquéfacteur* *réincorporateur ◑*	9.1.4-**100**
reversing heat exchanger	**intercambiador de calor para eliminación de condensables**	échangeur de chaleur réversible	9.1.4-**101**
rich liquid	**líquido rico**	liquide riche	9.1.4-**102**
roll-over effect	**efecto de desbordamiento**	effet de débordement	9.1.4-**103**
satellite station *vaporization station* *customer station*	**instalación de regasificación** instalación de vaporización	poste de vaporisation	9.1.4-**104**
send-out system *send-out unit*	**unidad de emisión**	unité d'émission	9.1.4-**105**

	ENGLISH	ESPAÑOL	FRANÇAIS
9.1.4-**106**	send-out unit *send-out system*	**unidad de emisión**	unité d'émission
9.1.4-**107**	shelf liquid ○ *poor liquid*	**líquido pobre**	liquide pauvre
9.1.4-**108**	side-arm column ● *argon column* *argon side column* ●	**columna de argón**	colonne d'argon
9.1.4-**109**	sieve tray	**bandeja-tamiz**	plateau perforé
9.1.4-**110**	single-column plant *Linde column* ●	**planta de columna unica**	unité à une seule colonne
9.1.4-**111**	sploshing	**bamboleo** vaivén	ballotement
9.1.4-**112**	stratification	**estratificación**	stratification
9.1.4-**113**	structured packing	**relleno estructurado**	garnissage structuré
9.1.4-**114**	submerged combustion vaporizer	**regasificador por combustión sumergida**	regazéifieur à combustion submergée *vaporisateur à combustion submergée*
9.1.4-**115**	sump	**cuba**	cuve
9.1.4-**116**	superinsulation *multilayer insulation (MLI)* *MLI*	**aislamiento reforzado**	superisolation
9.1.4-**117**	trumpler pass *balance stream* ●	**caudal de equilibrado** flujo de equilibrado	flux d'équilibrage
9.1.4-**118**	UHP GAN *ultra-high purity nitrogen (UHP GAN)*	**nitrógeno ultrapuro**	azote ultrapur
9.1.4-**119**	ultra-high purity nitrogen (UHP GAN) *UHP GAN*	**nitrógeno ultrapuro**	azote ultrapur
9.1.4-**120**	ultra-high purity oxygen	**oxígeno ultrapuro**	oxygène ultrapur
9.1.4-**121**	upper column ● *low-pressure column*	**columna de baja presión**	colonne basse pression
9.1.4-**122**	vacuum-insulated pipeline	**gasoducto aislado al vacío**	gazoduc isolé sous vide
9.1.4-**123**	vacuum-insulated tank *cryogenic storage tank*	**recipiente con aislamiento al vacío**	réservoir isolé sous vide
9.1.4-**124**	vaporization station *customer station* *satellite station*	**instalación de regasificación** instalación de vaporización	poste de vaporisation

	SECTION 9.2 *Cryobiology* **SECTION 9.2.1** *Cryobiology and the influence of low temperatures on living organisms*	**SUBCAPÍTULO 9.2** *Criobiología* **SUBCAPÍTULO 9.2.1** *Criobiología e influencia de las bajas temperaturas en los organismos vivos*	**SOUS-CHAPITRE 9.2** *Cryobiologie* **SOUS-CHAPITRE 9.2.1** *Cryobiologie et influence des basses températures sur les organismes vivants*
9.2.1-**1**	antifreeze protein	**proteina antihielo**	proteine antigel
9.2.1-**2**	biological material	**material biológico**	matériel biologique
9.2.1-**3**	biological tissue	**tejido biológico**	tissu biologique
9.2.1-**4**	bone marrow cell	**célula de médula ósea**	cellule de moelle osseuse

ENGLISH	ESPAÑOL	FRANÇAIS	
cellular rupture	**rotura celular**	éclatement des cellules	9.2.1-**5**
cellular structure	**estructura celular**	structure cellulaire	9.2.1-**6**
cold acclimatization	**aclimatación al frío**	acclimatation au froid	9.2.1-**7**
cold hardening	**resistencia al frío**	résistance au froid	9.2.1-**8**
cold preservation	**conservación en frío**	conservation à froid	9.2.1-**9**
cold-shock tolerance	**tolerancia al choque térmico frío**	tolérance au choc thermique à froid	9.2.1-**10**
cryo-etching	**criodecapado**	cryodécapage	9.2.1-**11**
cryobiology	**criobiología**	cryobiologie	9.2.1-**12**
cryobranding	**criomarcado** (subst.)	cryomarquage (des animaux)	9.2.1-**13**
cryoconservation *cryopreservation*	**crioconservación**	cryoconservation	9.2.1-**14**
cryopreservation *cryoconservation*	**crioconservación**	cryoconservation	9.2.1-**15**
cryoprotectant *cryoprotective agent* *cryoprotector* ○	**crioprotector**	cryoprotecteur *antigel biocompatible* ○	9.2.1-**16**
cryoprotective agent *cryoprotectant* *cryoprotector* ○	**crioprotector**	cryoprotecteur *antigel biocompatible* ○	9.2.1-**17**
cryoprotector ○ *cryoprotectant* *cryoprotective agent*	**crioprotector**	cryoprotecteur *antigel biocompatible* ○	9.2.1-**18**
crystal growth	**crecimiento de cristales**	croissance des cristaux	9.2.1-**19**
crystallization	**cristalización**	cristallisation	9.2.1-**20**
desiccation	**desecación**	dessiccation	9.2.1-**21**
devitrification	**desvitrificación**	dévitrification	9.2.1-**22**
differential scanning calorimetry	**calorimetría diferencial de barrido**	calorimétrie différentielle à balayage	9.2.1-**23**
dimethyl sulphoxide	**sulfóxido de dimetilo**	diméthyl sulfoxyde	9.2.1-**24**
dormancy	**letargo**	dormance	9.2.1-**25**
freeze tolerance	**tolerancia a la congelación**	tolérance à la congélation	9.2.1-**26**
frozen storage	**conservación en estado congelado**	conservation à l'état congelé	9.2.1-**27**
glycerol	**glicerol**	glycérol	9.2.1-**28**
heat-shock tolerance	**tolerancia al choque térmico**	tolérance au choc thermique	9.2.1-**29**
hibernation	**hibernación**	hibernation	9.2.1-**30**
hypothermal storage	**conservación en estado hipotérmico**	conservation en état hypothermique	9.2.1-**31**
ice nucleation	**nucleación del hielo**	nucléation de la glace	9.2.1-**32**
intracellular rehydration *intracellular resorption* ○	**resorción intracelular**	réhydratation intracellulaire *résorption intracellulaire* ○	9.2.1-**33**
intracellular resorption ○ *intracellular rehydration*	**resorción intracelular**	réhydratation intracellulaire *résorption intracellulaire* ○	9.2.1-**34**
lag period	**período de latencia**	période de latence	9.2.1-**35**
liposome	**liposoma**	liposome	9.2.1-**36**
low-temperature hazard	**peligros debidos a las bajas temperaturas** riesgos debidos a las bajas temperaturas	dangers des basses températures *risques dus aux basses températures* ○	9.2.1-**37**
low-temperature survival	**supervivencia a las bajas temperaturas**	survie aux basses températures	9.2.1-**38**

	ENGLISH	ESPAÑOL	FRANÇAIS
9.2.1-**39**	microbial flora *microflora* ○	**microflora** flora microbiana	flore microbienne *microflore* ○
9.2.1-**40**	microflora ○ *microbial flora*	**microflora** flora microbiana	flore microbienne *microflore* ○
9.2.1-**41**	osmolarity	**osmolaridad**	osmolarité
9.2.1-**42**	osmosis	**ósmosis**	osmose
9.2.1-**43**	osmotic shock	**choque osmótico**	choc osmotique
9.2.1-**44**	recrystallization	**recristalización**	recristallisation
9.2.1-**45**	rehydration	**rehidratación**	réhydratation
9.2.1-**46**	rewarming	**recalentamiento**	réchauffement
9.2.1-**47**	vernalization	**vernalización**	printanisation *vernalisation* ○
9.2.1-**48**	viability	**viabilidad**	viabilité
9.2.1-**49**	vitrification	**vitrificación**	vitrification
9.2.1-**50**	warming	**recalentamiento**	réchauffement
9.2.1-**51**	warming rate	**velocidad de recalentamiento**	vitesse de réchauffement

	SECTION 9.2.2 *Freeze-drying*	SUBCAPÍTULO 9.2.2 *Liofilización*	SOUS-CHAPITRE 9.2.2 *Lyophilisation*
9.2.2-**1**	atmospheric freeze-drying	**criodesecación atmosférica**	cryodessiccation atmosphérique
9.2.2-**2**	batch freeze-drying *discontinuous freeze-drying*	**liofilización discontinua**	lyophilisation discontinue *lyophilisation charge par charge* ○
9.2.2-**3**	centrifugal freeze-drying	**liofilización con centrifugación**	lyophilisation avec centrifugation
9.2.2-**4**	continuous freeze-drying	**liofilización continua**	lyophilisation continue
9.2.2-**5**	desiccation ratio	**relación de desecación**	taux de dessiccation
9.2.2-**6**	desorbable water ○ *releasable water*	**agua liberable**	eau libérable
9.2.2-**7**	diffuse sublimation front ○	**frente de sublimación difuso**	front de sublimation diffus ○
9.2.2-**8**	discontinuous freeze-drying *batch freeze-drying*	**liofilización discontinua**	lyophilisation discontinue *lyophilisation charge par charge* ○
9.2.2-**9**	drum freeze-drier	**liofilizador de tambor**	lyophilisateur à tambour
9.2.2-**10**	dryness ratio	**relación de sequedad** grado de sequedad	taux de siccité
9.2.2-**11**	excipient *freeze-drying additive*	**aditivo para liofilización** soporte para liofilización	support de lyophilisation *additif pour lyophilisation*
9.2.2-**12**	extraction ratio	**relación de extraccióna** grado de extracción	taux d'extraction
9.2.2-**13**	final temperature of freezing	**temperatura de completa solidificación**	température de congélation totale
9.2.2-**14**	freeze-dried *lyophilized*	**liofilizado** (adj.)	lyophilisé
9.2.2-**15**	freeze-drier	**liofilizador**	lyophilisateur

ENGLISH	ESPAÑOL	FRANÇAIS	
freeze-drying *lyophilization*	**liofilización** criodesecación	lyophilisation *cryodessiccation* ○	9.2.2-**16**
freeze-drying additive *excipient*	**aditivo para liofilización** soporte para liofilización	support de lyophilisation *additif pour lyophilisation*	9.2.2-**17**
lyophilic	**liófilo**	lyophile	9.2.2-**18**
lyophilizate	**liofilizado** (subst.)	lyophilisat	9.2.2-**19**
lyophilization *freeze-drying*	**liofilización** criodesecación	lyophilisation *cryodessiccation* ○	9.2.2-**20**
lyophilized *freeze-dried*	**liofilizado** (adj.)	lyophilisé	9.2.2-**21**
manifold drying apparatus	**cámara de vacío con conexiones múltiples**	hérisson	9.2.2-**22**
powder freezing	**congelación en polvo**	congélation en poudre	9.2.2-**23**
primary drying	**desecación primaria**	dessiccation primaire	9.2.2-**24**
rate of sublimation	**grado de sublimación**	taux de sublimation	9.2.2-**25**
releasable water *desorbable water* ○	**agua liberable**	eau libérable	9.2.2-**26**
residual moisture	**humedad residual**	humidité résiduelle	9.2.2-**27**
residual pressure	**presión residual**	pression résiduelle	9.2.2-**28**
secondary drying	**desecación secundaria**	dessiccation secondaire	9.2.2-**29**
sharp sublimation front	**frente de sublimación neto**	front net de sublimation *front aigu de sublimation* ○	9.2.2-**30**
shell freezing	**congelación en coquilla**	congélation en coquille	9.2.2-**31**
spray freeze-drying	**liofilización por pulverización**	lyophilisation par pulvérisation	9.2.2-**32**
sublimation front *sublimation interface*	**frente de sublimación** interfase de sublimación	front de sublimation *interface* ○	9.2.2-**33**
sublimation interface *sublimation front*	**frente de sublimación** interfase de sublimación	front de sublimation *interface* ○	9.2.2-**34**
sublimer	**sublimador**	sublimateur	9.2.2-**35**
tray drying chamber	**cámara de vacío con estanterías**	chambre à vide à étagères *chambre de dessiccation à étagères* ○	9.2.2-**36**
vacuum freezing	**congelación por vacío**	congélation par le vide	9.2.2-**37**
vibrating freeze-drier	**liofilizador de vibrador**	lyophilisateur à vibreur	9.2.2-**38**
water holding capacity	**capacidad de retención de agua**	capacité de rétention d'eau	9.2.2-**39**

SECTION 9.2.3 *Cryomedicine*	SUBCAPÍTULO 9.2.3 *Criomedicina*	SOUS-CHAPITRE 9.2.3 *Cryomédecine*	
(artificial) hibernation *hypothermia*	**hipotermia** hibernación artificial	hypothermie *hibernation (artificielle)* ○	9.2.3-**1**
biomaterial	**material biológico**	matériel biologique	9.2.3-**2**
blood bank	**banco de sangre**	banque de sang	9.2.3-**3**
bone bank	**banco de huesos**	banque d'os	9.2.3-**4**
cadaver storage *mortuary*	**depósito de cadáveres**	morgue	9.2.3-**5**

	ENGLISH	ESPAÑOL	FRANÇAIS
9.2.3-**6**	cooling cannula *cryogenic needle* *cryoprobe* *cryosurgical probe*	**criosonda (quirúrgica)**	cryosonde
9.2.3-**7**	cryo-ophthalmology	**criooftalmología**	cryo-ophtalmologie
9.2.3-**8**	cryoablation *cryoextirpation*	**crioablación** crioextirpación	cryoablation
9.2.3-**9**	cryoadherence *cryoadhesion*	**crioadherencia** crioadhesión	cryoadhérence
9.2.3-**10**	cryoadhesion *cryoadherence*	**crioadherencia** crioadhesión	cryoadhérence
9.2.3-**11**	cryoanalgesia	**crioanalgesia**	cryoanalgésie
9.2.3-**12**	cryoapplication	**crioaplicación**	cryoapplication
9.2.3-**13**	cryocauterization	**criocauterización**	cryocautérisation
9.2.3-**14**	cryocauterizer	**criocauterizador**	cryocautère
9.2.3-**15**	cryocoagulation	**criocoagulación**	cryocoagulation
9.2.3-**16**	cryodestruction	**criodestrucción**	cryodestruction
9.2.3-**17**	cryoextirpation *cryoablation*	**crioablación** crioextirpación	cryoablation
9.2.3-**18**	cryoextraction	**crioextracción**	cryoextraction
9.2.3-**19**	cryofixation *cryopexy* ○	**criopexia** criofijación	cryofixation *cryopexie* ○
9.2.3-**20**	cryogenic needle *cooling cannula* *cryoprobe* *cryosurgical probe*	**criosonda (quirúrgica)**	cryosonde
9.2.3-**21**	cryoimmunology	**crioinmunología**	cryoimmunologie
9.2.3-**22**	cryolesion	**criolesión**	cryolésion
9.2.3-**23**	cryomedicine	**criomedicina**	cryomédecine
9.2.3-**24**	cryometer	**criómetro**	cryothermomètre
9.2.3-**25**	cryopexy ○ *cryofixation*	**criopexia** criofijación	cryofixation *cryopexie* ○
9.2.3-**26**	cryoprobe *cooling cannula* *cryogenic needle* *cryosurgical probe*	**criosonda (quirúrgica)**	cryosonde
9.2.3-**27**	cryoresection	**criorresección**	cryorésection
9.2.3-**28**	cryoretinopexy	**criorretinopexia**	cryorétinopexie
9.2.3-**29**	cryoscalpel *cryostylet*	**criobisturí**	cryobistouri
9.2.3-**30**	cryostylet *cryoscalpel*	**criobisturí**	cryobistouri
9.2.3-**31**	cryosurgery	**criocirugía**	cryochirurgie
9.2.3-**32**	cryosurgical probe *cooling cannula* *cryogenic needle* *cryoprobe*	**criosonda (quirúrgica)**	cryosonde
9.2.3-**33**	cryotherapy *cryotreatment* ○	**crioterapia** criotratamiento	cryothérapie *cryotraitement* ○

ENGLISH	ESPAÑOL	FRANÇAIS	
cryotool	**crioinstrumental (quirúrgico)**	cryo-outil	9.2.3-**34**
cryotreatment ○ *cryotherapy*	**crioterapia** criotratamiento	cryothérapie *cryotraitement* ○	9.2.3-**35**
graft *transplant*	**injerto**	greffon	9.2.3-**36**
hypothermia *(artificial) hibernation*	**hipotermia** hibernación artificial	hypothermie *hibernation (artificielle)* ○	9.2.3-**37**
hypothermic blanket	**manta hipotérmica**	couverture réfrigérante	9.2.3-**38**
mortuary *cadaver storage*	**depósito de cadáveres**	morgue	9.2.3-**39**
tissue bank	**banco de tejidos**	banque de tissus	9.2.3-**40**
transplant *graft*	**injerto**	greffon	9.2.3-**41**
transplantation	**trasplante**	transplantation	9.2.3-**42**
University of Wisconsin solution	**solución UW para conservación**	solution de conservation UW	9.2.3-**43**

Capítulo 10.

OTRAS APLICACIONES DEL FRÍO

- ◑ término aceptado
- ○ término obsoleto

ENGLISH	ESPAÑOL	FRANÇAIS	
SECTION 10.1 *Water ice*	**SUBCAPÍTULO 10.1** *Hielo hídrico*	**SOUS-CHAPITRE 10.1** *Glace hydrique*	
block ice *cake ice* ◑	**hielo en barras** hielo en bloques	glace en bloc *glace en pain* *glace en mouleaux* ◑	10.1-**1**
briquette ice ○	**hielo en briquetas**	glace en briquettes ○	10.1-**2**
broken ice *lump ice* ◑	**hielo troceado**	glace concassée	10.1-**3**
cake ice ◑ *block ice*	**hielo en barras** hielo en bloques	glace en bloc *glace en pain* *glace en mouleaux* ◑	10.1-**4**
chip ice *chipped ice* ◑	**hielo en laminillas**	glace en copeaux	10.1-**5**
chipped ice ◑ *chip ice*	**hielo en laminillas**	glace en copeaux	10.1-**6**
clear ice	**hielo transparente**	glace transparente	10.1-**7**
crushed ice	**hielo triturado**	glace broyée	10.1-**8**
crystal ice	**hielo cristalino**	glace cristal	10.1-**9**
cube ice	**hielo in cubitos**	glace en cubes	10.1-**10**
flake ice *slice ice* *scale ice* ◑	**hielo en copos** hielo en radajas hielo en escamas	glace en écailles *glace en éclats* ◑ *glace en flocons* ◑	10.1-**11**
ice (to)	**suministrar hielo**	glacer	10.1-**12**
ice block *ice cake* ◑	**barra de hielo** bloque de hielo	bloc de glace *pain de glace* ◑ *mouleau de glace* ◑	10.1-**13**
ice cake ◑ *ice block*	**barra de hielo** bloque de hielo	bloc de glace *pain de glace* ◑ *mouleau de glace* ◑	10.1-**14**
ice cube	**cubito de hielo**	cube de glace *glaçon*	10.1-**15**
lump ice ◑ *broken ice*	**hielo troceado**	glace concassée	10.1-**16**
opaque ice *white ice*	**hielo opaco**	glace opaque	10.1-**17**
plate ice	**hielo en placas**	glace en plaques	10.1-**18**
processed ice	**hielo fraccionado**	glace fractionnée	10.1-**19**
ribbon ice	**hielo en cintas**	glace en ruban	10.1-**20**
scale ice ◑ *flake ice* *slice ice*	**hielo en copos** hielo en radajas hielo en escamas	glace en écailles *glace en éclats* ◑ *glace en flocons* ◑	10.1-**21**
sea-water ice	**hielo de agua de mar**	glace d'eau de mer	10.1-**22**
shell ice ○ *tube ice*	**hielo en tubos**	glace en tubes	10.1-**23**
sized ice	**hielo calibrado**	glace calibrée	10.1-**24**
slice ice *flake ice* *scale ice* ◑	**hielo en copos** hielo en radajas hielo en escamas	glace en écailles *glace en éclats* ◑ *glace en flocons* ◑	10.1-**25**
slush	**nieve fundente**	neige fondante	10.1-**26**

	ENGLISH	ESPAÑOL	FRANÇAIS
10.1-**27**	slush ice	**hielo nieve humedecido**	glace-neige mouillée
10.1-**28**	small ice	**hielo dividido**	glace divisée
10.1-**29**	snow	**nieve**	neige
10.1-**30**	snow-ice	**hielo nieve**	glace-neige
10.1-**31**	tube ice *shell ice* ○	**hielo en tubos**	glace en tubes
10.1-**32**	(water) ice	**hielo (hídrico)**	glace (hydrique)
10.1-**33**	white ice *opaque ice*	**hielo opaco**	glace opaque

	SECTION 10.2 *Ice-making plants*	SUBCAPÍTULO 10.2 *Fábricas de hielo*	SOUS-CHAPITRE 10.2 *Fabriques de glace*
10.2-**1**	air agitation (in ice making)	**insuflación de aire (en los moldes de hielo)**	insufflation d'air (dans les mouleaux à glace)
10.2-**2**	brine tank ○ *ice-making tank* *ice tank* ○	**tanque generador de hielo** tanque de salmuera	bac à glace *bac à saumure* *bac générateur de glace*
10.2-**3**	can dump (USA) ○ *ice tip*	**basculador de moldes de hielo** volcador de moldes de hielo	basculeur de mouleaux à glace *culbuteur de mouleaux à glace*
10.2-**4**	can filler	**llenador (de moldes)**	emplisseur (de mouleaux) *herse de remplissage* ○ *remplisseur (de mouleaux)* ○
10.2-**5**	core pulling and filling system	**dispositivo de extracción y llenado del núcleo**	dispositif de succion du noyau
10.2-**6**	dip tank ○ *thawing tank*	**tanque de desmoldeo**	bac de démoulage
10.2-**7**	freezing cylinder ○ *ice generator* *freezing drum* ○	**equipo generador de hielo** organo generador de hielo	organe générateur de glace
10.2-**8**	freezing drum ○ *ice generator* *freezing cylinder* ○	**equipo generador de hielo** organo generador de hielo	organe générateur de glace
10.2-**9**	ice can ○ *ice mould*	**molde para hielo**	mouleau (à glace)
10.2-**10**	ice can frame ○ *ice mould frame* *ice can grid* ○	**armadura para moldes** bastidor para moldes	châssis pour mouleaux
10.2-**11**	ice can grid ○ *ice mould frame* *ice can frame* ○	**armadura para moldes** bastidor para moldes	châssis pour mouleaux
10.2-**12**	ice can group ○ *row of moulds* *row of cans* ○	**batería de moldes**	rangée de mouleaux
10.2-**13**	ice can truck	**mecanismo móvil para moldes**	treuil mobile pour mouleaux
10.2-**14**	ice cellar *snow cellar* ○	**almacén de hielo**	glacière
10.2-**15**	ice chute	**tobogán para hielo**	glissière à glace
10.2-**16**	ice crane	**puenta grúa para tanque de hielo**	pont roulant de bac à glace

ENGLISH	ESPAÑOL	FRANÇAIS	
ice crusher	**trituradora de hielo**	broyeur à glace	10.2-**17**
ice dump table	**mesa de desmoldeo**	table de démoulage	10.2-**18**
ice factory ● *ice-making plant*	**fábrica de hielo**	fabrique de glace	10.2-**19**
ice generator *freezing cylinder* ● *freezing drum* ●	**equipo generador de hielo** organo generador de hielo	organe générateur de glace	10.2-**20**
ice-maker	**generador de hielo**	générateur de glace	10.2-**21**
ice-making plant *ice factory* ●	**fábrica de hielo**	fabrique de glace	10.2-**22**
ice-making tank *brine tank* ● *ice tank* ●	**tanque generador de hielo** tanque de salmuera	bac à glace *bac à saumure* *bac générateur de glace*	10.2-**23**
ice manufacture	**fabricación de hielo**	fabrication de la glace	10.2-**24**
ice mould *ice can* ●	**molde para hielo**	mouleau (à glace)	10.2-**25**
ice mould frame *ice can frame* ● *ice can grid* ●	**armadura para moldes** bastidor para moldes	châssis pour mouleaux	10.2-**26**
ice storage room	**cámara de almacenamiento de hielo**	réserve à glace *resserre à glace* *glacière* ●	10.2-**27**
ice tank ● *ice-making tank* *brine tank* ●	**tanque generador de hielo** tanque de salmuera	bac à glace *bac à saumure* *bac générateur de glace*	10.2-**28**
ice tip *can dump (USA)* ●	**basculador de moldes de hielo** volcador de moldes de hielo	basculeur de mouleaux à glace *culbuteur de mouleaux à glace*	10.2-**29**
row of cans ● *row of moulds* *ice can group* ●	**batería de moldes**	rangée de mouleaux	10.2-**30**
row of moulds *ice can group* ● *row of cans* ●	**batería de moldes**	rangée de mouleaux	10.2-**31**
snow cellar ● *ice cellar*	**almacén de hielo**	glacière	10.2-**32**
thawing tank *dip tank* ●	**tanque de desmoldeo**	bac de démoulage	10.2-**33**
water forecooler ● *water precooler*	**preenfriador de agua**	prérefroidisseur d'eau	10.2-**34**
water precooler *water forecooler* ●	**preenfriador de agua**	prérefroidisseur d'eau	10.2-**35**

SECTION 10.3 *Winter sports*	SUBCAPÍTULO 10.3 *Deportes de invierno*	SOUS-CHAPITRE 10.3 *Sports d'hiver*	
curling rink	**pista de "curling"**	piste de curling	10.3-**1**
de-icing	**fusión de la pista de hielo**	fusion du plateau de glace	10.3-**2**
ice hockey rink	**pista de patinaje para hockey sobre hielo**	patinoire pour hockey	10.3-**3**
ice rink ● *skating rink*	**pista de patinaje**	patinoire	10.3-**4**

	ENGLISH	ESPAÑOL	FRANÇAIS
10.3-**5**	Ice slab	**pista de hielo**	plateau de glace
10.3-**6**	olympic rink	**pista olímpica de patinaje**	patinoire olympique
10.3-**7**	pipe (cooling) grids	**red de tubos (enfriadores)**	réseau de tubes (refroidisseurs)
10.3-**8**	rink floor	**pavimento soporte de la pista** forjado soporte de la pista	soubassement de patinoire
10.3-**9**	skating rink *ice rink* ❍	**pista de patinaje**	patinoire
10.3-**10**	snow gun *snow maker*	**proyector de nieve**	canon à neige *enneigeur* ❍
10.3-**11**	snow maker *snow gun*	**proyector de nieve**	canon à neige *enneigeur* ❍
10.3-**12**	speed track	**pista de velocidad**	anneau de vitesse
10.3-**13**	surfacing (of ice)	**acabado (del hielo)**	surfaçage (de la glace)

	SECTION 10.4 *Chemical industries*	SUBCAPÍTULO 10.4 *Industria química*	SOUS-CHAPITRE 10.4 *Industries chimiques*
10.4-**1**	brackish water	**agua salobre**	eau saumâtre
10.4-**2**	cryoconcentration *freeze-concentration* ❍	**concentración por congelación** crioconcentración	cryoconcentration *concentration par congélation* ❍
10.4-**3**	cryogrinding *freeze-grinding* ❍	**criotrituración**	cryobroyage
10.4-**4**	cryotrimming *cryotumbling* ❍	**criodesbarbado**	cryoébarbage
10.4-**5**	cryotumbling ❍ *cryotrimming*	**criodesbarbado**	cryoébarbage
10.4-**6**	freeze-concentration ❍ *cryoconcentration*	**concentración por congelación** crioconcentración	cryoconcentration *concentration par congélation* ❍
10.4-**7**	freeze-grinding ❍ *cryogrinding*	**criotrituración**	cryobroyage
10.4-**8**	freeze desalination	**desalinización por congelación**	dessalement par congélation
10.4-**9**	freeze out (to)	**separar por congelación**	séparer par congélation
10.4-**10**	soft water	**agua dulce**	eau douce

	SECTION 10.5 *Metallurgy and mechanical industries*	SUBCAPÍTULO 10.5 *Industrias metalúrgica y mecánica*	SOUS-CHAPITRE 10.5 *Industries métallurgiques et mécaniques*
10.5-**1**	ageing (of materials)	**envejecimiento (de los materiales)**	vieillissement (des matériaux)
10.5-**2**	cold brittleness	**fragilidad al frío**	fragilité au froid
10.5-**3**	cold-dimensional stabilization	**estabilización dimensional por el frío**	stabilisation dimensionnelle par le froid
10.5-**4**	cold-shrink fitting *expansion fitting* ❍	**machihembrado por frío**	assemblage par contraction *emmanchement à froid*
10.5-**5**	cryohardening	**criotemple**	cryotrempe

ENGLISH	ESPAÑOL	FRANÇAIS	
expansion fitting ● *cold shrink fitting*	**machihembrado por frío**	assemblage par contraction *emmanchement à froid*	10.5-**6**
hardening *quenching* ●	**temple**	trempe	10.5-**7**
quenching ● *hardening*	**temple**	trempe	10.5-**8**
shrink disassembly	**desmontaje por contracción**	démontage par contraction	10.5-**9**

SECTION 10.6 *Miscellaneous applications of refrigeration*	SUBCAPÍTULO 10.6 *Aplicaciones diversas del frío*	SOUS-CHAPITRE 10.6 *Applications diverses du froid*	
aggregate cooling	**enfriamiento de los agregados**	refroidissement des agrégats	10.6-**1**
concrete-dam cooling	**enfriamiento de presas de hormigón**	refroidissement des barrages en béton	10.6-**2**
embedded cooling coils	**tubos de enfriamiento encastrados**	tubes de refroidissement noyés	10.6-**3**
freezing the soil	**congelación del suelo** congelación del terreno	congélation du sol *congélation du terrain* ●	10.6-**4**
frozen earth storage ● *frozen ground storage* *frozen soil storage* ●	**almacenamiento en terreno congelado**	stockage en terrain congelé *stockage en terre gelée* ●	10.6-**5**
frozen ground storage *frozen earth storage* ● *frozen soil storage* ●	**almacenamiento en terreno congelado**	stockage en terrain congelé *stockage en terre gelée* ●	10.6-**6**
frozen soil storage ● *frozen ground storage* *frozen earth storage* ●	**almacenamiento en terreno congelado**	stockage en terrain congelé *stockage en terre gelée* ●	10.6-**7**
permafrost	**profundidad de congelación permanente (del terreno)** permafrost	pergélisol *permagel* ●	10.6-**8**
simulation chamber	**cámara de simulación**	chambre de simulation	10.6-**9**
space simulator	**simulador espacial**	simulateur spatial	10.6-**10**
thermal shroud	**envolvente de protección térmica**	enveloppe de protection thermique	10.6-**11**

Capítulo 11.

FRÍO Y MEDIO AMBIENTE

◐ término aceptado

○ término obsoleto

ENGLISH	ESPAÑOL	FRANÇAIS	
CHAPTER 11 *Refrigeration and the environment*	**CAPÍTULO 11** *Frío y medio ambiente*	**CHAPITRE 11** *Froid et environnement*	
Article 5 country	**Estado Artículo 5**	pays de l'Article 5	**11-1**
atmospheric lifetime	**permanencia en la atmósfera**	durée de vie atmosphérique	**11-2**
climate change	**cambio climático**	changement climatique	**11-3**
climate change attributed to human activity	**cambio climático debido a la actividad humana** cambio climático antropogénico	changement climatique attribué à l'activité humaine	**11-4**
Conference of the Parties *COP*	**Conferencia de las Partes**	Conférence des Parties	**11-5**
controlled substance	**sustancia regulada**	substance réglementée *substance sous contrôle* ◐	**11-6**
COP *Conference of the Parties*	**Conferencia de las Partes**	Conférence des Parties	**11-7**
direct warming impact	**efecto directo de calentamiento** efecto directo del efecto invernadero	impact direct sur le réchauffement planétaire *impact climatique*	**11-8**
drop-in replacement activity	**fluido frigorígeno para sustitución inmediata**	remplacement immédiat *drop-in* ◐	**11-9**
fluorocarbon	**fluorocarburo**	fluorocarbure	**11-10**
global environmental issues	**temas medioambientales globales**	enjeux climatiques planétaires	**11-11**
global warming *greenhouse effect*	**calentamiento global** efecto invernadero	réchauffement planétaire *effet de serre*	**11-12**
global warming potential *GWP*	**potencial de calentamiento global** potencial de efecto invernadero	potentiel de réchauffement planétaire *GWP*	**11-13**
greenhouse effect *global warming*	**calentamiento global** efecto invernadero	réchauffement planétaire *effet de serre*	**11-14**
greenhouse gas	**gas de efecto invernadero**	gaz à effet de serre	**11-15**
GWP *global warming potential*	**potencial de calentamiento global** potencial de efecto invernadero	potentiel de réchauffement planétaire *GWP*	**11-16**
halocarbon	**halocarburo**	hydrocarbure halogéné *halocarbure*	**11-17**
halon	**halón**	halon	**11-18**
indirect warming impact	**efecto indirecto de calentamiento** efecto indirecto del efecto invernadero	impact indirect sur le réchauffement planétaire	**11-19**
Kyoto Protocol	**Protocolo de Kyoto**	Protocole de Kyoto	**11-20**
life-cycle cost analysis	**análisis del coste del ciclo de vida**	analyse du coût (d'une installation) pendant son cycle de vie	**11-21**
Life-Cycle Climate Performance (LCCP)	**comportamiento climático durante el ciclo de vida**	impact sur le climat au cours du cycle de vie	**11-22**
Montreal Protocol	**Protocolo de Montreal**	Protocole de Montréal	**11-23**
ODP *ozone depletion potential*	**ODP (potencial de agotamiento del ozono)** ODP (potencial de agotamiento del ozono estratosférico) ODP (potencial de agotamiento de la capa de ozono)	potentiel d'appauvrissement de la couche d'ozone *ODP*	**11-24**

	ENGLISH	ESPAÑOL	FRANÇAIS
11-25	ODS *ozone-depleting substance*	**ODS (sustancia que agota el ozono)** ODS (sustancia que agota el ozono estratosférico) ODS (sustancia que agota la capa de ozono)	substance appauvrissant l'ozone
11-26	ozone	**ozono**	ozone
11-27	ozone-depleting substance *ODS*	**ODS (sustancia que agota el ozono)** ODS (sustancia que agota el ozono estratosférico) ODS (sustancia que agota la capa de ozono)	substance appauvrissant l'ozone
11-28	ozone depletion	**agotamiento del ozono** agotamiento del ozono estratosférico agotamiento de la capa de ozono	appauvrissement de l'ozone
11-29	ozone depletion potential *ODP*	**ODP (potencial de agotamiento del ozono)** ODP (potencial de agotamiento del ozono estratosférico) ODP (potencial de agotamiento de la capa de ozono)	potentiel d'appauvrissement de la couche d'ozone *ODP*
11-30	ozone layer	**capa de ozono**	couche d'ozone
11-31	phase-out	**eliminación escalonada (de una sustancia)**	élimination d'ozone
11-32	radiative forcing	**potencial radiativo** influencia potencial sobre el equilibrio radiante terrestre	forçage radiatif
11-33	retrofit	**reconversión**	conversion
11-34	stratosphere	**estratosfera**	stratosphère
11-35	TEWI *Total Equivalent Warming Impact*	**TEWI (equivalente del efecto total de calentamiento)** TEWI (equivalente del efecto invernadero total)	TEWI
11-36	Total Equivalent Warming Impact *TEWI*	**TEWI (equivalente del efecto total de calentamiento)** TEWI (equivalente del efecto invernadero total)	TEWI
11-37	transitional substance	**sustancias de transición**	frigorigène de transition
11-38	ultraviolet radiation	**radiación ultravioleta**	rayonnement ultraviolet

ÍNDICE ALFABÉTICO: ESPAÑOL

línea de líquido   2.6.4-**40**
línea de salmuera   2.6.4-**3**
linea isoentálpica   1.2.3-**24**
lineal de expositores   5.2-**33**
línea límite de carga   5.2-**35**
lineal multitemperatura de expositores   5.2-**60**
liofilización   9.2.2-**16**
liofilización con centrifugación   9.2.2-**3**
liofilización continua   9.2.2-**4**
liofilización discontinua   9.2.2-**2**
liofilización por pulverización   9.2.2-**32**
liofilizado (adj.)   9.2.2-**14**
liofilizador   9.2.2-**15**
liofilizador de tambor   9.2.2-**9**
liofilizador de vibrador   9.2.2-**38**
liofilizado (subst.)   9.2.2-**19**
liófilo   9.2.2-**18**
liposoma   9.2.1-**36**
líquido   2.9.1-**13**
líquido criogénico   9.1.1-**23**
líquido incongelable   2.9.3.1-**4**
líquido pobre   9.1.4-**98**
líquido rico   9.1.4-**102**
líquido saturado   1.2.2-**55**
líquido subenfriado   1.2.2-**62**
lira de dilatación   2.6.4-**18**
listones para estibar   3.2-**11**
llave   2.6.1-**19**   2.6.1-**89**
llave angular   2.6.1-**2**
llave con bridas   2.6.1-**26**
llave de brida única   2.6.1-**74**
llave de carga   2.6.1-**17**
llave de carga de aceite   2.6.1-**50**
llave de cierre   2.6.1-**73**
llave de compuerta   2.6.1-**32**
llave de extracción   2.6.1-**79**
llave de paso   2.6.1-**73**
llave de purga   2.6.1-**29**   2.6.1-**61**
llave de purga de aceite   2.6.1-**51**
llave de retención   2.6.1-**40**
llave de servicio   2.6.1-**69**
llave de vaciado   2.6.1-**29**
llave general   2.6.1-**41**
llave manual de paso   2.6.1-**34**
llave para botella de fluido frigorígeno   2.6.1-**23**
llave principal   2.6.1-**41**
llave sin empaquetadura   2.6.1-**54**
llenador (de moldes)   10.2-**4**
local habitado   3.1-**6**
longitud aleteada   2.5.1-**10**
longitud de coherencia   9.1.3-**3**
longitud de la escala   1.1.7-**63**
longitud útil del tubo aleteado   2.5.1-**35**
lote   6.5-**3**
lubricación   3.5.2-**13**
lubricación a presión   3.5.2-**12**
lubricación forzada   3.5.2-**12**
lubricación por barboteo   3.5.2-**5**
lucha contra los olores   6.4-**19**
lumbrera de reducción de potencia   2.4.3-**54**
luminancia   1.3.2-**19**
luminaria ventilada   7.2.1-**41**

M

machihembrado por frío   10.5-**4**
macropartícula   7.7-**12**
maduración   6.3.1-**42**
maduración acelerada   6.3.1-**2**
maduración (de las carnes)   6.3.2-**2**
maduración de quesos   6.3.3-**15**
madurez fisiológica   6.3.1-**39**
magnesia   3.3.2-**53**
magnitud adimensional   1.1.7-**19**
magnitud de influencia   1.1.7-**28**
magnitud derivada   1.1.7-**13**
magnitud de salida   1.1.7-**45**

magnitud fundamental   1.1.7-**6**
magnitud influyente   1.1.7-**28**
magnitud (medible)   1.1.7-**34**
magnitud regulada   1.1.7-**11**
magnitud reguladora   1.1.7-**12**
mal olor   7.2.3-**18**
mampara   5.5.4-**3**
mampara pantalla   5.6-**3**
mando   2.8-**14**
mando a distancia   2.8-**92**
mando automático   2.8-**4**
manga   5.3-**23**
mangueta de cigüeñal   2.4.3-**18**
manguito de unión   2.6.4-**14**
manipulación   5.6-**15**
manipulación con tarimas   5.6-**31**
manipular con tarimas   5.6-**32**
manómetro   1.1.3-**38**   3.5.1-**35**
manómetro compuesto   1.1.3-**9**
manómetro de aguja   1.1.3-**23**
manómetro de alta (presión)   2.7-**13**
manómetro de aspiración   2.7-**37**
manómetro de baja (presión)   2.7-**37**
manómetro de Bourdon   1.1.3-**8**
manómetro de columna de líquido   1.1.3-**28**
manómetro de glicerina   1.1.3-**26**
manómetro de impulsión   2.7-**13**
manómetro de ionización   1.1.3-**25**
manómetro de membrana   1.1.3-**13**
manómetro de montador   1.1.3-**44**
manómetro de presión de aceite   3.5.2-**23**
manómetro de seguridad   1.1.3-**43**
manómetro de vacío   1.1.3-**56**
manómetro de viscosidad   1.1.3-**60**
manómetro duplex   1.1.3-**16**
manómetro térmico   1.1.3-**50**
manovacuómetro   1.1.3-**9**
manta   3.3.2-**2**
manta aislante   3.3.1-**2**   3.3.1-**18**
manta con armadura   3.3.3-**37**
manta hipotérmica   9.2.3-**38**
máquina enfriadora   7.1-**14**
máquina frigorífica   2.3.1-**9**
máquina (frigorífica) de absorción   2.3.3.1-**5**
máquina frigorífica de absorción discontinua   2.3.3.1-**20**
máquina frigorífica de absorción intermitente   2.3.3.1-**20**
máquina (frigorífica) de ciclo de aire   2.3.3.4-**1**
máquina frigorífica de eyección (de vapor)   2.3.3.2-**3**
marchitamiento   6.2-**73**
martilleo (de una válvula de expansión)   3.5.1-**54**
masa "directa"   6.3.7-**8**
masa específico   1.1.1-**9**
masa molecular relativa   1.1.1-**32**
masa "semidirecta"   6.3.7-**6**
masa semipreparada   6.3.7-**6**
masa terminada   6.3.7-**8**
masilla aislante   3.3.3-**19**
masilla de estanqueidad   3.3.4-**24**
mat   3.3.2-**2**
matadero   6.3.2-**1**
material aislante   3.3.2-**48**
material aislante térmico   3.3.2-**80**
material anticorrosivo   6.2-**18**
material biológico   9.2.1-**2**   9.2.3-**2**
material celular   3.3.2-**10**
material criogénico   9.1.1-**21**
material no absorbente   6.2-**52**
material no tóxico   6.2-**54**
material para junta   2.6.4-**30**
mecánico frigorista   1.1.1-**29**
mecanismo de accionamiento   2.6.1-**53**
mecanismo de álabes de prerrotación   2.4.4-**21**
mecanismo de la válvula   2.6.1-**88**
mecanismo móvil para moldes   10.2-**13**

medición   1.1.7-**38**
medida   1.1.7-**36**   1.1.7-**38**
medidas de control   6.2-**14**
medidor de flujo de calor   1.3.1-**7**
medidor de presión   1.1.3-**23**   1.1.7-**47**
medidor de presión absoluta   1.1.3-**2**
medidor de presión con elemento de medida elástico   1.1.3-**24**
medidor de presión diferencial   1.1.3-**15**
medidor de presión para control automático   1.1.3-**11**
medidor de presión para encastrar   1.1.3-**19**
medidor de presión para panel   1.1.3-**19**
medio congelador   2.9.3.1-**22**
medio dispersante   7.3.1-**28**
medio exterior al sistema termodinámico   1.2.1-**63**
mercancías no enfriadas   5.5.4-**11**
mesa de desmoldeo   10.2-**18**
mesa refrigerada para pastelería   6.3.7-**1**
meseta de congelación (en un diagrama de temperatura-tiempo)   4.3-**32**
metanero de cubas prismáticas   9.1.4-**60**
metanero del tipo membrana   9.1.4-**58**
metanero del tipo Moss   9.1.4-**59**
metano   2.9.2.5-**18**
metanol   2.9.3.3-**15**
método de congelación   4.3-**30**
método de enfriamiento   4.1-**4**
método de medida   1.1.7-**42**
método (de recuperación) aspiración-compresión   2.9.1-**5**
metrología   1.1.7-**43**
mezcla   1.1.6-**58**   2.9.1-**3**
mezcla azeotrópica   1.2.2-**2**
mezcla congeladora   2.9.3.1-**23**
mezcla de refrigerantes   9.1.4-**76**
mezcla enfriadora   2.9.3.1-**18**
mezcla eutéctica   2.9.3.1-**21**   2.9.3.3-**10**
mezcla líquido-vapor   1.2.2-**43**
mezcla no-azeotrópica   1.2.2-**47**
mezcla para helados   6.3.4-**13**
mezcla refrigerante   2.9.3.1-**18**
mezcla zeotrópica   1.2.2-**47**
microclima   7.1-**45**
microflora   9.2.1-**39**
mirilla   2.7-**56**   3.2-**38**
mirilla de inspección   3.2-**29**
miscibilidad   1.1.6-**56**
modelo de dos fluidos   9.1.2-**19**
modulante   2.8-**37**
módulo termoeléctrico   2.3.3.3-**7**
molde para hielo   10.2-**25**
moldura   3.3.3-**25**
mondongo   6.3.2-**25**
monitorización   2.8-**63**
montador frigorista   1.1.1-**28**
montaje empotrado   4.1-**15**
motocompresor hermético accesible   2.4.1-**14**
motor de expansión   9.1.4-**30**
mueble con frente accesible   5.2-**51**
mueble conservador (de helados)   6.3.4-**8**
mueble con servicio posterior   5.2-**50**
mueble de endurecimiento (de helados)   6.3.4-**10**
mueble exposición   5.2-**14**
mueble expositor   5.2-**14**
mueble frigorífico   5.2-**43**
mueble frigorífico de media altura   5.2-**55**
mueble frigorífico vertical con estantes   5.2-**61**
mueble mostrador   5.2-**4**
muelle   5.4-**22**
multisplit   7.5-**27**
multizona   7.1-**46**
muro cortina   3.2-**9**
muro de bloques huecos   3.2-**6**
muro de carga   3.2-**17**
muro de contención   9.1.4-**13**
muro de doble pared con cámara de aire   3.2-**7**

válvula de seccionamiento 2.6.1-**36**
válvula de seguridad 2.8-**84**
válvula de seguridad de contrapeso 2.8-**20**
válvula de seguridad de resorte 2.8-**114**
válvula de servicio 2.6.1-**69**
válvula de servicio y control 2.6.1-**62**
válvula de solenoide 2.6.2-**13**
válvula de tres vías 2.6.1-**81**
válvula de vacío 2.6.1-**60**
válvula de vástago 2.4.3-**61** 2.4.3-**88**
válvula diversora 2.6.1-**25**
válvula electromagnética 2.6.2-**13**
válvula esférica 2.6.1-**6** 2.6.1-**57**
válvula esférica de salida múltiple 2.6.1-**45**
válvula hidráulica 2.6.2-**11**
válvula inversora 2.6.1-**66**
válvula limitadora de presión 2.5.2-**6** 2.8-**82** 2.8-**84**
válvula manual de cierre 2.6.1-**34**
válvula mezcladora 2.6.1-**43**
válvula moduladora 2.6.2-**14**
válvula motorizada 2.6.2-**15**
válvula neumática 2.6.2-**19**
valvula neumatica de arranque 3.5.1-**81**
válvula obturadora 2.6.1-**58**
válvula piloto 2.6.2-**18** 2.6.3-**23**
válvula presostática 2.6.2-**21**
válvula reductora de presión 2.6.2-**22**
válvula reguladora de presión 2.6.2-**20**
válvula servoaccionada 2.6.2-**17**
válvula servocomandada 2.6.2-**17**
válvula sin empaquetadura 2.6.1-**54**
válvula termostática 2.6.2-**28**
vapor 1.2.2-**74**
vapor de agua 1.1.4-**66**
vapor de escape 2.3.3.1-**18**
vapor húmedo 1.2.2-**75**
vapor instantáneo 3.5.1-**24**
vaporización 1.2.2-**73**
vaporización instantánea 3.5.1-**25**
vaporizador de ambiente 9.1.4-**2**
vaporizado (subst.) 9.1.4-**9**
vapor saturado (seco) 1.2.2-**26**
vapor sobrecalentado 1.2.2-**69**
vapor sobresaturado 1.2.2-**70**
variación de la potencia (de un compresor) 3.5.1-**8**
variación de temperatura 1.1.2-**66**
variante de corazón pardo 6.3.1-**15**
varilla de fijación aislante 3.3.3-**5**
vaso Dewar 9.1.1-**54**
vástago de válvula 2.4.3-**96**
vatio (W) 1.1.1-**43**
vehículo compartimentalizado 5.5.2-**12**
vehículo de transporte terrestre 5.5.1-**20**
vehículo frigorífico 5.5.1-**21**
vehículo isotermo 5.5.1-**17**
vehículo isotermo de reparto 5.5.2-**3**
vehículo refrigerado 5.5.1-**27**
vehículo refrigerante 5.5.1-**27**
vehículo ventilado 5.5.1-**29**
velocidad axial 7.2.2-**17**
velocidad crítica 1.1.6-**15**
velocidad crítica (He II) 9.1.2-**2**
velocidad de aire 7.2.2-**13**
velocidad de arrastre 9.1.2-**1**
velocidad de congelación 4.3-**34**
velocidad de corriente 1.1.6-**98**
velocidad de enfriamiento 4.2-**4**

velocidad de extracción de calor 2.2-**12**
velocidad de flujo 1.1.6-**98**
velocidad del sonido 1.1.6-**75**
velocidad de recalentamiento 9.2.1-**51**
velocidad de regulación 2.8-**17**
velocidad frontal 7.2.2-**37**
velocidad másica 1.1.6-**51**
velocidad terminal 7.2.2-**59**
ventanillo de aireación 3.3.4-**5**
ventanillo de inspección 3.2-**29**
ventilación 7.2.2-**61**
ventilación del espacio interior de las cerchas 7.2.2-**16**
ventilación forzada 7.2.2-**38**
ventilación forzada controlada 7.4-**11**
ventilación forzada controlada de doble flujo 7.4-**13**
ventilación forzada controlada de simple flujo 7.4-**53**
ventilación mecánica controlada 7.4-**11**
ventilación mecánica controlada de doble flujo 7.4-**13**
ventilación mecánica controlada de simple flujo 7.4-**53**
ventilación nocturna 7.5-**28**
ventilación por falso techo 3.2-**50**
ventilación transversal 7.2.2-**22**
ventilador 7.4-**20**
ventilador alta temperatura 7.4-**29**
ventilador antideflagrante 7.4-**30**
ventilador axial 7.4-**2**
ventilador axial con guías 7.4-**61**
ventilador axial de baja presión 7.4-**2**
ventilador axial de envolvente 7.4-**57**
ventilador axial montado en pletina 7.4-**47**
ventilador bifurcado 7.4-**4**
ventilador bi-rotatorio 7.4-**10**
ventilador centrífugo 7.4-**8**
ventilador centrífugo-helicoidal 7.4-**43**
ventilador con conducto(s) 7.4-**18**
ventilador de alabes aerodinámicas 7.4-**1**
ventilador de álabes directores 7.4-**28**
ventilador de alta presión 7.4-**6**
ventilador de aspiración 7.4-**38**
ventilador de doble admisión 7.4-**14**
ventilador de impulsión 7.4-**6**
ventilador de palas aerodinámicas 7.4-**1**
ventilador de pared 7.4-**46**
ventilador de recirculación 7.4-**9**
ventilador de tejado 7.4-**48**
ventilador de turbina eólica 5.5.1-**1**
ventilador en anillo 7.4-**52**
ventilador en chorro 7.4-**42**
ventilador en línea 7.4-**58**
ventilador en montaje hermético 7.4-**26**
ventilador helicoidal 7.4-**2**
ventilador multi-etapa 7.4-**44**
ventilador para gases húmedos 7.4-**64**
ventilador para usos industriales 7.4-**39**
ventilador reversible 7.4-**51**
ventilador tangencial 7.4-**12**
ventiloconvector 7.5-**16**
ventiloconvector para calefacción 7.3.2-**4**
venturi 1.1.6-**100**
verificación 6.2-**84**
vermiculita 3.3.2-**85**
vernalización 9.2.1-**47**
viabilidad 9.2.1-**48**

vibración acústica 1.4.1-**1**
vibración del álabe 2.4.4-**3**
vida potencial de almacenamiento 6.1-**28**
vidrio celular 3.3.2-**8**
viga 3.2-**4**
viga fría 7.5-**10**
vigilancia 3.5.1-**52** 6.2-**51**
vigilancia de calidad 6.2-**64**
viscosidad 1.1.6-**103**
viscosidad cinemática 1.1.6-**47**
viscosidad dinámica 1.1.6-**23**
viscosímetro 1.1.6-**101**
visor 2.7-**56**
visor de líquido 2.7-**31**
visor de nivel de aceite 3.5.2-**21**
vitrificación 9.2.1-**49**
vitrina 5.2-**14**
vitrina de autoservicio 5.2-**53**
vitrina frigorífica 5.2-**45**
vitrina para productos congelados 5.2-**23**
volante 3.5.1-**27**
volcador de moldes de hielo 10.2-**29**
voltaje especificado 1.4.2-**4**
volumen barrido 2.4.3-**57**
volumen bruto de una cámara frigorífica 5.3-**14**
volumen bruto nominal 5.1-**34**
volumen bruto total 5.1-**15**
volumen bruto total nominal 5.1-**36**
volumen crítico 1.2.2-**21**
volumen de almacenamiento 5.2-**59**
volumen desplazado 2.4.3-**57**
volumen desplazado real 2.4.3-**1**
volumen desplazado teórico 2.4.3-**84** 2.4.3-**85**
volumen específico 1.1.1-**39**
volumen específico del aire húmedo (referido al aire seco) 1.1.4-**34**
volumen interior bruto 5.1-**27** 5.2-**27**
volumen interior neto 5.2-**28**
volumen interno 2.5.1-**59**
volumen límite de carga (capacidad) 5.1-**27**
volumen másico 1.1.1-**39**
volumen másico del aire húmedo 1.1.4-**34**
volumen total de almacenamiento 5.1-**16**
volumen total de una cámara frigorífica 5.3-**14**
volumen total útil 5.1-**16**
volumen útil de una cámara frigorífica 5.3-**18**
volumen útil nominal 5.1-**35**
volumen útil total nominal 5.1-**37**
voluta 2.4.4-**23**

W

"winterización" 6.3.7-**9**

Z

zeolita 2.9.4-**6**
zeótropo 1.2.2-**47**
zona de preparación de pedidos 5.4-**19**
zona de respiración 7.2.3-**6**
zona de riesgo 7.7-**19**
zona limpia 7.7-**5**
zona mínima de propagación 9.1.3-**25**
zona muerta 7.2.2-**24**
zona neutra 2.8-**24**
zona ocupada 7.1-**49**
zonificación 7.1-**65**